William Stuart Auchincloss

The Practical Application of the Slide Valve and Link Motion to Stationary, Portable, Locomotive and Marine Engines

With New and Simple Methods for Proportioning the Parts

William Stuart Auchincloss

The Practical Application of the Slide Valve and Link Motion to Stationary, Portable, Locomotive and Marine Engines
With New and Simple Methods for Proportioning the Parts

ISBN/EAN: 9783337420178

Printed in Europe, USA, Canada, Australia, Japan

Cover: Foto ©berggeist007 / pixelio.de

More available books at **www.hansebooks.com**

THE

PRACTICAL APPLICATION

OF THE

SLIDE VALVE AND LINK MOTION

TO

STATIONARY, PORTABLE, LOCOMOTIVE, AND MARINE ENGINES,

WITH

NEW AND SIMPLE METHODS

FOR

PROPORTIONING THE PARTS.

BY

WILLIAM S. AUCHINCLOSS, C. E.,
M. AMER. SOC. C. E.

XIIITH EDITION,
REVISED.

NEW YORK:

D. VAN NOSTRAND COMPANY,
23 MURRAY, AND 27 WARREN STS.

1895.

PREFACE.

LINK AND VALVE MOTIONS has had a phenomenal sale during the past twenty-five years. It has proved itself both a standard authority with Mechanical Engineers and Draughtsmen, and a valued text-book with Colleges and Technical Schools. Its market has not been confined to the United States, but it has found ready sale in Great Britain. It was so favorably considered by the noted journal—"Engineering, of London"—that it closed its critical review of the book with these words :

> "All the matters we have mentioned are treated
> with a clearness and absence of unnecessary verbi-
> age, which renders the work a peculiarly valuable
> one. The TRAVEL SCALE only requires to be known
> to be appreciated. Mr. A. writes so ably on his
> subject, we wish he had written more."

About ten years ago, Julius Springer, of Berlin, published Link and Valve Motions in German under the title: SCHIEBER UND COULISSENSTEURUNGEN, edited by Herr A. Müller, Chief Engineer of the Borsig Locomotive Works.

In the present Edition the Author has carefully eliminated all abstruse formulæ, because he considers it absurd to invoke the aid of higher mathematics for the solution of everyday problems in Link and Valve Motion. The component parts of such motions are always compact and the distances small, consequently they do not involve such delicate angles, arcs, sines, cosines and tangents as in Astronomy, and should not be so treated, but all dimensions should be computed either arithmetically or graphically by the most simple and direct processes.

He is deeply sensible of the generous reception accorded his Work by the Profession, and since the book deals exclusively with fundamental principles (to the neglect of patented devices), he sends it forth anew, confident that in its revised form it will prove specially acceptable to all Engineering students and practical Machinists who appreciate quick short-hand methods.

W. S. A.

March. 1895.

CONTENTS.

—

PART I.

PART II.

PART III.

PART IV.

PART V.

TRAVEL SCALE.

PART I.

THE SLIDE VALVE.

ELEMENTARY PRINCIPLES

AND

GENERAL PROPORTIONS.

POWER AND WORK.

THE fundamental query in designing a steam-engine has reference to the power required to accomplish a given amount of work.

The term *work*, when employed in a mathematical sense, signifies the continuous overcoming of an offered resistance along a definite path.

The *quantity of work* is the product of that resistance into the space passed over.

As the standards of weight and distance differ throughout the world, the expressions for quantity of work also differ. With the English standard of pounds avoirdupois and feet, the quantity of work is said to consist of a certain number of *foot-pounds*. But with the French standard of weight, the kilogramme ($=2.20462$ lbs. avoirdupois) and of distance, the mètre ($=3.28089$ ft.), the expression becomes a certain number of *kilogrammètres*.

Thus the quantity of work expended in raising a weight of 300 lbs. through a vertical height of 10 ft. $=3,000$ ft.-lbs. and that of elevating a weight of 50 kilogrammes to a height of 20 mètres $=1.000$ kilogrammètres. The quantity of work performed by the steam in the cylinder of an engine, equals the mean effective pressure exerted upon the entire area of the piston multiplied by the space passed over in a given time. The interval of time usually taken is *one min-ute;* hence, if the distance traveled by the piston during a single revolution of the crank be multiplied by the number of revolutions made per minute, their product will equal the required space.

Suppose, for instance, the mean effective pressure on each square inch of a piston, having an area of 1,500 sq. ins., is 60 lbs. : then the total pressure will be 1,500 × 60 = 90,000 lbs., and if the crank makes 40 revolutions per minute, with a piston stroke of 3 ft., the speed of the piston becomes 3 ft. × 2 × 40 = 240 ft. per minute ; consequently the quantity of work = 90,000 lbs. × 240 ft. = 21,600,000 ft.-lbs.

1.—HORSE POWER.

A force capable of raising a weight of 33,000 lbs. one foot high in one minute is termed a Horse power.

The expression originated at the time of the discovery of the steam-engine from the necessity which then arose for comparing its powers with those of the prevailing motor. In its early history this unit had three prefixes—Nominal, Indicated, and Actual—derived from the various methods of estimating the power. The nominal horse power was based on the general practice of the age, which dealt with low pressures and slow piston speeds. These quantities have of late years been greatly increased and the old formula in consequence, grown of less and less importance as a true expression of relative capacity.

Indicated horse power designates the total unbalanced power of an engine employed in overcoming the combined resistances of friction and the load. Hence it equals the quantity of work performed by the steam in one minute, divided by 33,000. Thus, in the above example, the indicated horse power equals

$$\frac{21,600,000}{33,000} = 11)\overline{7,200} \quad \begin{array}{l} 3)21,600 \\ \overline{654 \text{ HP}} \end{array}$$

The mean effective pressure can alone be determined by means of an instrument called the Indicator.

The Actual or net horse power, expresses the total available power of an engine, hence it equals the indicated horse power less an amount expended in overcoming the friction. The latter has two components, viz: the power required to run the engine, detached from its load, at the normal speed, and that required when it is connected with its load. It is customary in designing *massive engines*—in the absence of reliable data—to estimate the loss of available pressure by the unloaded friction at 2 lbs. per square inch, and subsequently to deduct 7½ per cent. for the friction of the load. Thus, if the mean pressure of the steam within the cylinder... = 60 ^{lbs. per sq. in.}

It becomes 58 after allowing for unloaded friction, 58
And 7½ % of this for the friction of the load....... = 4.4
    Gives a net pressure of.................... 53.6 lbs.

But for *small engines* of the ordinary design the total loss by friction will, in many instances, amount to 15 or 20 % of the mean pressure.

Thus, if the mean pressure...................... = 60 lbs.
15 % of 60 = total loss by friction .... = 9 "
    Gives an available pressure of............ 51 "

The French apply the term *Force de cheval* to a power capable of raising 4.500 kilogrammes 1 mètre high in 1 minute. Reducing these quantities to their equivalents in pounds and feet and multiplying together, we find that their horse-power equals a force capable of raising 32,549 lbs. 1 foot high in a minute, which is about $\frac{1}{13}$ less than the English unit of measure.

The following TABLE furnishes the Force de cheval equivalents of horse powers ranging between 10 and 100 :

Horse Power.	Force de cheval.	Horse Power.	Force de cheval.
10	10.14	60	60.83
15	15.20	65	65.89
20	20.28	70	70.97
25	25.34	75	76.03
30	30.41	80	81.11
35	35.48	85	86.17
40	40.55	90	91.25
45	45.62	95	96.31
50	50.69	100	101.3856
55	55.75	. .	. . .

For powers greater than 100, and less than 1,000, multiply these terms by 10 ; or, if in excess of 1,000, multiply by 100.

II.—MEAN EFFECTIVE PRESSURE.

The character of the connections between the boiler and steam cylinder, their length, degree of protection, number of bends, shape of valves, etc., must all be considered in forming an estimate of the initial steam pressure in the cylinder ; while the mean effective pressure will depend upon the point of cut-off of the steam, and the freedom with which it exhausts.

The exact portion of the stroke that should be completed before this closure or cut-off takes place is a vexed question among engineers, and its discussion is foreign to the object of this Treatise, in which—with the exception of noting certain limits prescribed by different valve motions—it will be considered as predetermined.

Having chosen a point of cut-off, and having estimated the initial pressure of the steam for a given boiler pressure, the question of mean pressure exerted by the steam throughout the piston's stroke, can be approximately solved by the subjoined *Table*, which has been computed in the ordi-

Mean Pressure, Volume, and Temperature Table.

Initial Pressure	Temperature Fah.	Relative Volume	STROKE = 1. MEAN PRESSURE FOR VARIOUS CUT-OFFS.						
			¼ or 0.25	⅜ or 0.375	½ or 0.5	⅝ or 0.625	⅔ or 0.666	¾ or 0.75	⅞ or 0.875
Lbs.	Deg.		Lbs.			Lbs.			Lbs.
20	260	765	11.9	14.9	16.9	18.4	18.7	19.3	19.8
25	267	677	14.9	18.6	21.2	23.	23.3	24.1	24.7
30	274	608	17.9	23.3	25.4	27.6	28.	28.9	29.7
35	281	552	20.9	26.	29.6	32.1	32.7	33.7	34.6
40	287	506	23.9	29.7	33.9	36.8	37.3	38.5	39.6
45	293	467	26.8	33.4	38.1	41.3	42.	43.4	44.5
50	298	434	29.8	37.1	42.3	45.9	46.7	48.2	49.5
55	303	406	32.8	40.8	46.6	50.5	51.3	53.	54.4
60	308	381	35.8	44.5	50.8	55.1	56.	57.8	59.4
65	312	359	38.8	48.2	55.	59.7	60.7	62.6	64.3
70	316	340	41.7	52.	59.3	64.3	65.3	67.5	69.3
75	320	323	44.7	55.7	63.5	68.9	69.9	72.3	74.2
80	324	307	47.7	59.4	67.7	73.5	74.6	77.1	79.2
85	328	293	50.7	63.1	71.9	78.1	79.3	81.9	84.1
90	332	281	53.7	66.8	76.2	82.7	84.	86.7	89.1
95	335	269	56.7	70.5	80.4	87.3	88.7	91.6	94.
100	338	259	59.7	74.2	84.6	91.9	93.3	96.4	99.
105	341	249	62.6	77.9	88.9	96.5	97.9	101.1	103.9
110	344	239	65.6	81.6	93.1	101.1	101.6	105.9	108.9
115	347	231	68.6	85.3	97.4	105.6	106.3	110.8	113.8
120	350	223	71.6	89.	101.6	110.2	110.9	115.6	118.8
125	353	216	74.6	92.7	105.8	114.8	115.6	120.5	123.7
130	356	209	77.6	96.4	110.	119.4	120.3	125.3	128.7
135	358	203	80.6	100.1	114.2	124.	125.	130.1	133.6
140	360	197	83.5	103.8	118.5	128.6	130.6	134.9	138.6
145	363	191	86.5	107.5	122.7	133.2	135.3	139.7	143.5
150	365	186	89.5	111.2	126.9	137.8	140.	144.5	148.5
Common difference			3.0	3.7	4.2	4.6	4.7	4.8	5.0

nary manner with the aid of logarithms (Naperian Base). The first column is given for pressures above that of the atmosphere, or the same as registered by an ordinary steam-gauge. The second and third, for temperature and volume, are taken from Mons. Regnault's Experiments on Saturated Steam. In the estimate for volume, that of the water producing the steam was considered equal to Unity. The Table makes no allowance for clearance.

If from the mean pressure we subtract the mean value of the back pressure, or that which may arise from imperfections in the exhaust, which is usually taken for low-pressure engines at from 1 to 2 lbs. per square inch, the resulting pressure will be the mean effective pressure (in pounds) exerted on each square inch of the piston and may be represented by the letter P.

For high-pressure engines (having an ordinary slide valve) a more exact determination of the mean effective pressure may be secured from the subjoined table, which embodies the results of 50 experiments made by Mr. Gooch, in 1851, with the locomotive "Great Britain," whose boiler pressure varied from 60 to 150 lbs. per square inch.

Mean Effective Pressures incident to a Simple Slide-Valve Motion for various Cut-offs.

Cut-Off at—	Mean Pressure. (Boiler press. = 1.00.)	Cut-Off at—	Mean Pressure. (Boiler press. = 1.00.)
0.1	0.15	0.45	0.62
$0.125 = \frac{1}{8}$	0.2	$0.5 = \frac{1}{2}$	0.67
0.15	0.24	0.55	0.72
0.175	0.28	$0.625 = \frac{5}{8}$	0.79
0.2	0.32	$0.666 = \frac{2}{3}$	0.82
$0.25 = \frac{1}{4}$	0.4	0.7	0.85
0.3	0.46	$0.75 = \frac{3}{4}$	0.89
$0.333 = \frac{1}{3}$	$0.5 = \frac{1}{2}$	0.8	0.93
$0.375 = \frac{3}{8}$	0.55	$0.875 = \frac{7}{8}$	0.98
0.4	0.57	. . . .	. . .

Given { Boiler pressure = 70 lbs. per sq. in.
{ Steam cut off at $\frac{2}{3}$ of the stroke.

Required.—The mean effective pressure P ?

We learn from the table that this pressure for a cut-off of $\frac{2}{3}$ the stroke is 0.82 of the boiler pressure.

Then $70 \times 0.82 = 57.4$, or

The mean effective pressure P $= 57.4$ lbs. per sq. in.

III.—SPEED OF PISTON.

The speed S, or number of feet travelled by the piston in one minute, like the subject of cut-off, rests with the judgment of the individual designer. Nothing more will be attempted in this connection than the presentation of quantities most frequently found in ordinary practice :

Small stationary engines from.......170 to 230 ft. per min.
Large stationary engines...........250 to 300   "
(Rarely as high as 350 ft.)
River and Sound steamer engines....350 to 500   "
Marine engines....................250 to 600   "
The Corliss stationary engine.......400 to 500   "
(Usually 50 revolutions.)
Locomotive engines about...........600   "
(Occasionally 700 or 800 ft.)
The Allen engine...................600 to 800   "
(Generally the former speed.)

It is interesting to note that a fine specimen of the latter

2

form of engine was operated successfully by Mr. Charles T. Porter, during the late "Exposition Universelle," at the astonishing speed of 1,400 feet per minute.

IV.—DIAMETER OF PISTON.

Having decided the questions relating to indicated horse power, mean available pressure P and piston speed S, all the elements are at hand for determining the area of the piston, and consequently its diameter.

The formula for indicated horse power, solved with reference to such area, will read :

$$A = \frac{33,000 \times \text{Horse power}}{S \times P}$$

or, Area of piston is found by *multiplying the required indicated horse power by 33,000, and dividing the product by speed of piston multiplied by the mean available pressure.*

The corresponding diameter can be obtained from an Area Table.

EXAMPLE.

Suppose that the indicated horse power=100.
Piston speed=300 ft. per minute.
Mean available pressure=21 lbs.
Then the

$$\text{Area} = \frac{33,000 \times 100}{300 \times 21} = 523.8 \text{ sq. in.}$$

Which gives a diameter of about 26 inches.

V.—STROKE OF PISTON.

The general expression for the stroke of an engine (in feet) is,

$$\text{Stroke} = \frac{\text{Piston Speed}}{2 \times \text{No. of Revolutions}}.$$

conversely,

$$\text{No. of Revolutions} = \frac{\text{Piston Speed}}{2 \times \text{Stroke}}.$$

There are many circumstances tending to limit the stroke of a piston. Among other considerations the diameter of a paddle-wheel influences the number of revolutions that can advantageously be made by the crank of a side-wheel steamer, and consequently determines the stroke when the piston speed is chosen. Peculiarities of design frequently make it desirable that an engine should be run at a slow speed and transmit its power through gearing.

Again, the diameters of pulleys for shafting exert an influence, as when the main shaft of a shop is required to run at 120 revolutions per minute, then 60 revolutions for the crank of the engine, will allow a ratio of 2 : 1 between the diameter of the band wheel and shaft pulley.

With a very rapid piston speed, the stroke of the engine is due more to a length imposed on the connecting rod by the necessities of the design, than to the number of revolutions of the crank. In the case of the locomotive, the stroke is generally about 24 inches, and the piston speed 600 feet per minute, while the speed of the engine which depends on its power and the diameter of its drivers, ranges between 20 and 60 miles per hour.

The accompanying table has been calculated, for drivers of different diameters, to represent the number of revolutions they will make per minute, irrespective of slip, when the engine travels at given speeds per hour.

Revolutions made by Driving Wheels of Locomotive at given speeds.

Driving-wheel diameter.	SPEED IN MILES PER HOUR.						Revolutions per mile.
	20 miles.	25.	30.	35.	40.	50.	
4 ft. 0 in.	140	175	210	..	..		420.
4 " 3 "	132	165	198	..	..		395.5
4 " 6 "	124	156	186	..	..		373.6
4 " 9 "	118	148	177	207	..		354.
5 " 0 "		140	168	196	..		336.
5 " 3 "		134	160	187	..		320.2
5 " 6 "		128	153	179	204		305.9
5 " 9 "		..	146	170	195		292.3
6 " 0 "		..	140	163	187		280.3
6 " 3 "		..	135	157	179	224	269.
6 " 6 "		..	129	150	172	216	258.6
7 " 0 "		..	120	140	160	200	240.

The subjoined table is applicable to stationary and marine engines:

No. of Revolutions of Crank for Given Stroke and (approximate) Piston Speed.

Stroke.	PISTON SPEED.														
	Ft. 200	210	220	225	230	240	250	260	270	280	290	300	320	340	Ft. 350
1 ft. 6 in.	67	70	73	75	76	80	83	86	90	93	97	100	106	113	116
1 " 8 "	60	63	66	68	70	72	75	78	81	84	87	90	96	100	105
1 "10 "	55	57	60	61	63	66	68	71	74	76	79	82	88	93	96
2 " 0 "	50	52	55	56	57	60	63	65	67	70	72	75	80	85	87
2 " 3 "	44	47	49	50	51	53	55	58	60	62	64	66	72	76	78
2 " 6 "	40	42	44	45	46	48	50	52	54	56	58	60	64	68	70
2 " 9 "	36	38	40	41	42	43	45	47	49	51	53	55	58	62	64
3 " 0 "	33	35	36	37	38	40	42	43	45	47	48	50	53	56	58
3 " 3 "	31	32	33	34	35	37	38	40	41	43	44	46	50	52	54
3 " 6 "	29	30	31	32	33	34	36	37	38	40	41	43	46	48	50
3 " 9 "	27	28	29	30	31	32	33	34	36	37	39	40	43	45	47
4 " 0 "	25	26	27	28	29	30	31	32	34	35	36	38	40	42	44
4 " 3 "	23	24	25	26	27	28	29	30	32	33	34	35	38	40	41
4 " 6 "	22	23	24	25	26	27	28	29	30	31	32	33	35	38	39
4 " 9 "	21	22	23	23	24	25	26	27	28	29	30	31	33	36	37
5 " 0 "	20	21	22	22	23	24	25	26	27	28	29	30	32	34	35

These dimensions, the stroke of piston and diameter of cylinder, are so constantly used in comparing engines of different powers, that, as far as possible, they should consist of whole numbers quite free from all fractions of an inch.

VI.—AREA OF STEAM PORT.

This dimension ranks next to cut-off in its controlling influence upon the proportions of the valve seat and face. It may justly be considered as a *Base* from which all the other dimensions are derived in conformity with certain laws. Its value depends greatly upon the manner in which the port is employed, whether simply for admitting the steam to the cylinder, or for purposes both of admission and exit. In cases of admission it is evident that the pressure will be sustained at substantially a constant quantity by the flow of steam from the boiler. But in cases of exit or exhaust, a limited quantity of steam, impelled by a constantly *diminishing* pressure, forces its way into the atmosphere with less and less velocity. If, then, the engine is supplied with two steam and two exhaust passages, the ports will be correctly proportioned when the areas of the latter *exceed* those of the former by an amount indicated by careful experiment. When, however, one passage performs *both* duties, it should have an area suitable for the exhaust and be opened only a limited amount for the admission of the steam. Very excellent results have been found to attend the employment of an area equal to 0.04 of that of the piston, and a steam-pipe area of 0.025 of the same, when the speed of the piston does not exceed 200 ft.

per minute, but widely-different factors are demanded by higher speeds, like those peculiar to locomotives.

In the year 1844 M. M. Gouin and Le Chatelier instituted a series of experiments for ascertaining the value of such terms. These were continued about six years later by Messrs. Clark, Gooch, and Bertera, upon engines of British manufacture. The various results having been collated and analyzed by Mr. Clark, were finally presented to the public in his valuable work on "Railway Locomotives." From this it appears that with a piston speed of 600 ft. per minute, an area of 0.1 that of the piston was found to give practically a perfect exhaust, a steam-pipe area of 0.08 a free admission of steam to the chest, and a port opening of from 0.6 to 0.9 the entire width of the port, depending on the humidity of the steam, a free admission to the cylinder.

The following table has been prepared for intermediate speeds of the piston on the assumption that for average lengths of pipe the area increases as the speed, and that a higher speed is usually attended by increased pressure:

Speed of Piston.	Port Area.	Steam-Pipe Area.
200 feet per minute.	.04 area of piston.	.025 area of piston.
250 " "	.047 " "	.032 " "
300 " "	.055 " "	.039 " "
350 " "	.062 " "	.046 " "
400 " "	.07 " "	.053 " "
450 " "	.077 " "	.06 " "
500 " "	.085 " "	.067 " "
550 " "	.092 " "	.074 " "
600 " "	.1 " "	.08 " "

Having determined the area of the steam port, the next step will be to resolve it into its factors, length and breadth. When a small travel of the valve is essential, the length should be made as nearly equal to the diameter of the cylinder as possible : then the port area divided by the length, furnishes of course the value of the breadth or S in Fig. 1.

The extent to which the valve should open this port for the admission of the steam will equal from 0.6 to 0.9 of the value of S. and the *minimum travel* of the valve, that which with a given cut-off just opens the steam port the amount of this limit. The *maximum travel* is governed by expediency, the general tendency of an excess over the minimum travel is to render the events of the stroke *more decisive*, the cut-off takes place with greater brevity, avoiding unnecessary wire drawing of the steam and the release opens rapidly, affording a more perfect exit. Where the travel is small, these good qualities should be secured by increasing the travel, until the valve gives an opening equal to or even greater than the width of the steam port. With a large travel no such attempt should be made, since it would inevitably sacrifice much work in friction and cause a far greater loss than gain.

EXAMPLE.

Diameter of a certain piston=26 inches. Area=531.

Piston speed=350 ft. per minute.

Required.—Width of steam port, minimum width of port opening and diameter of supply steam pipe.

From the Tables we have :

Sq. inches.

Area of steam port=531 × .062=33 sq. inches.

The length of the port=diameter of cylinder=26″.

And the width=$\frac{33}{26}$=1.3 inches or $1\frac{5}{16}$.

Minimum width of port opening=0.6 × 1.3=$\frac{3}{4}$ inch.

Sq. inches.

Area of steam pipe=531 × .046=24.4 sq. inches.

Consequently the diameter=$5\frac{1}{2}$ inches.

In the Corliss Engine, where the steam is admitted and exhausted through different valves, it is customary to give the steam passage an area of $\frac{1}{15}$ to $\frac{1}{16}$ that of the piston, and the exhaust an area of from $\frac{1}{10}$ to $\frac{1}{11}$.

In this connection a few remarks may appropriately be
made with reference to the formation of the valve edge and
the walls of the steam port. The experiments of scientists
like Weisbach, D'Aubuisson and Koch, prove that the various phenomena of contraction in the fluid vein observed in
the flow of water are equally true for gases, the formulæ of
discharge however have slightly different coefficients of
efflux. The character of the discharge will evidently vary
with the extent of opening offered by the valve edge, from
what is termed "discharge through a thin plate" at the
commencement, to that through a "short tube" with the
full opening. Fig. 1 illustrates the natural convergence

Fig. 1.

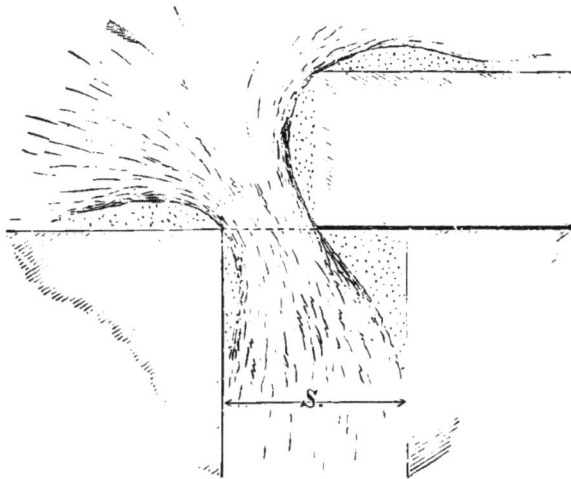

which takes place in the filaments of the steam vein with
the common slide valve. If the edge were formed as in
Fig. 2 the discharge would be much improved and rendered similar to that which occurs through an ordinary
"mouth piece."

The curvature of the valve edge should commence far

Fig. 2.

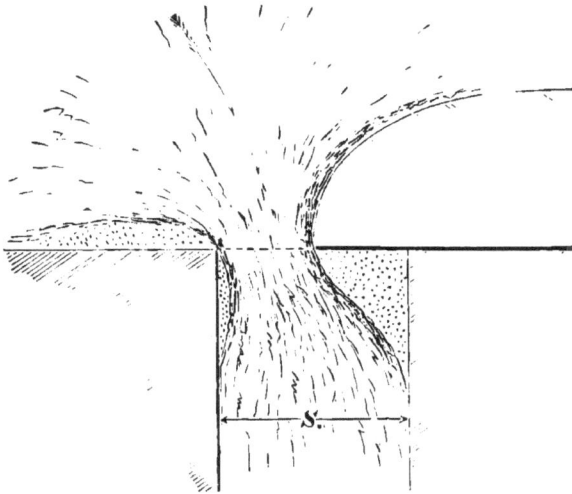

enough above the rubbing surfaces to permit a limited amount of wear without altering the proportion of the parts.

Every effort should also be made to reduce the amount of clearance for the steam and loss of head by friction, to a minimum value. Hence the passage from the port to the cylinder must be constructed as short as possible, be of uniform cross section and bend with easy curves if bending is indispensable.

In the moulding of a cylinder casting, the cores for the steam and exhaust passages should be faced with very great care, in order to secure surfaces along which the steam will flow with perfect freedom.

PISTON, CRANK

AND

VALVE MOTIONS.

In essaying the study of an intricate subject like the relative motions of the piston and the ordinary slide valve of a steam engine, it is of the utmost importance to first divest the parts of all the complicating influences which arise from special constructions and present them in such simple and elementary forms, that the discovery of the fundamental laws governing their motions may be facilitated. If these are clearly defined, the deduction of others adapted to special cases will subsequently be accomplished with comparative ease.

The entire series of events which take place within the cylinder of an engine, occur when the piston has reached definite positions in its complete stroke. It follows (since there is in practice no fixed limit to the stroke) that an infinite number of such positions may be occupied, and in order to express them by a standard which shall apply equally to all cases, a unit scale must be adopted. The stroke of all pistons therefore will be regarded throughout this Treatise as equal to *Unity*, and their positions at certain important periods, as decimal portions of the entire stroke.

If a movable point is caused to travel around a fixed

one, in the same plane, at a constant distance therefrom, it will describe a curved line called a circle. For the purpose of locating any position in the path of the movable point, the circle has from remote ages—though not wisely— been divided into 360 equal parts called degrees (360°), each degree into 60 minutes and each minute into 60 seconds.

While the piston of an engine performs a single stroke, the crank-pin makes a semi-revolution (180°) about the centre of the main shaft, each position of the former conse- quently corresponds with some angular position of the crank-arm, and if these angles are arranged in a Table we can instantly determine therefrom the number of degrees over which the pin must pass in order to bring the piston to any desired position.

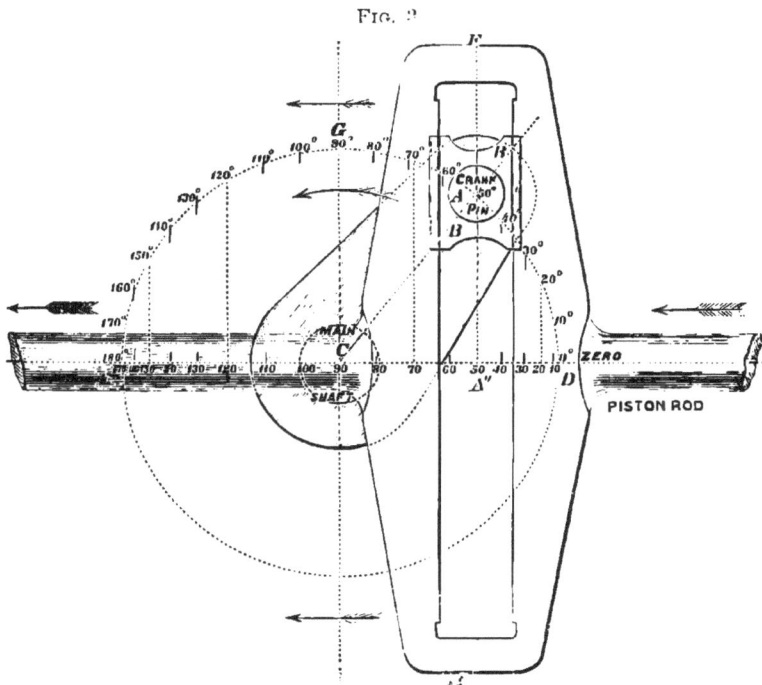

Fig. 2

Since the "slotted cross-head" shown in Fig. 3 is the only form of connection between the crank-pin and piston, in which the piston moves from one extremity of the stroke to the other at the same speed as the crank-pin—measured on the stroke line—it will answer our purpose for determining the fundamental principles of the piston and valve motions. The arrangement of the parts are clearly shown in the Figure. The crank-pin is surrounded by blocks BB, these slide freely up and down the solid frame FH to which the piston-rod is welded, so that while the crank-pin advances from D to G the block mounts towards F, returns as it approaches E and descends towards H on the return stroke ED. For convenience, the cylinder will always be regarded as lying on the *right-hand* side of the main shaft and the point of the crank-pin circle *nearest* to the cylinder as the zero or starting point of the forward stroke.

TABLE A.

Piston Position.	Crank Angle.	Piston Position.	Crank Angle.	Piston Position.	Crank Angle.
	Deg.		Deg.		Deg.
0.1	$36\frac{7}{8}$	$0.5625 = \frac{9}{16}$	$97\frac{1}{8}$	$0813 = \frac{13}{16}$	$128\frac{3}{8}$
$0.125 = \frac{1}{8}$	$41\frac{3}{8}$	0.575	$98\frac{5}{8}$	0.82	$129\frac{3}{8}$
0.15	$45\frac{5}{8}$	0.6	$101\frac{1}{8}$	0.83	$131\frac{1}{8}$
0.175	$49\frac{1}{2}$	$0.625 = \frac{5}{8}$	$104\frac{1}{8}$	0.84	$132\frac{7}{8}$
0.2	$53\frac{1}{8}$	0.65	$107\frac{1}{8}$	0.85	$134\frac{3}{8}$
0.225	$56\frac{5}{8}$	$0.666 = \frac{2}{3}$	$109\frac{1}{8}$	0.86	$136\frac{1}{8}$
$0.25 = \frac{1}{4}$	60	0.68	$111\frac{5}{8}$	0.87	$137\frac{3}{8}$
0.275	$63\frac{1}{8}$	$0.687 = \frac{11}{16}$	112	$0.875 = \frac{7}{8}$	$138\frac{5}{8}$
0.3	$66\frac{3}{8}$	0.69	$112\frac{3}{8}$	0.88	$139\frac{5}{8}$
0.325	$69\frac{1}{8}$	0.7	$113\frac{5}{8}$	0.89	$141\frac{1}{8}$
$0.333 = \frac{1}{3}$	$70\frac{1}{8}$	0.71	$114\frac{7}{8}$	0.9	$143\frac{1}{8}$
0.35	$72\frac{1}{8}$	0.72	$116\frac{1}{8}$	0.91	$145\frac{5}{8}$
$0.375 = \frac{3}{8}$	$75\frac{5}{8}$	0.73	$117\frac{3}{8}$	0.92	$147\frac{1}{8}$
0.4	$78\frac{1}{8}$	0.74	$118\frac{5}{8}$	0.93	$149\frac{3}{8}$
0.425	$81\frac{3}{8}$	$0.75 = \frac{3}{4}$	120	0.94	$151\frac{5}{8}$
$0.437 = \frac{7}{16}$	$82\frac{7}{8}$	0.76	$121\frac{3}{8}$	0.95	$154\frac{1}{8}$
0.45	$84\frac{1}{8}$	0.77	$122\frac{5}{8}$	0.96	$156\frac{5}{8}$
0.475	$87\frac{1}{8}$	0.78	$124\frac{1}{8}$	0.97	$160\frac{1}{8}$
$0.5 = \frac{1}{2}$	90	0.79	$125\frac{1}{8}$	0.98	$163\frac{3}{8}$
0.525	$92\frac{7}{8}$	0.8	$126\frac{7}{8}$	0.99	$168\frac{1}{8}$
0.55	$95\frac{3}{8}$	0.81	$128\frac{1}{8}$	1.00	180

The foregoing Table furnishes angular positions of the crank-arm corresponding with the various points in the stroke which may at times be occupied by the piston.

To illustrate its application, suppose for—

EXAMPLE.

The stroke of a certain piston=36 inches.

Query.—How many degrees will the crank have passed over when the piston reaches points respectively 9″ and 23⅜″ distant from the commencement of its stroke?

1st. $\dfrac{9''}{36''}=$ $6)\overline{9.00}$ $=6)\overline{1.50}=0.25$ of the stroke.
   0.25

2d. $\dfrac{23\frac{3}{8}}{36}=\dfrac{23.38}{36}=0.649$ of the stroke.

Then by the Table:

0.25 of the stroke=an angular passage of 60°.
0.65 "   =   "   "   107½°
The required angles.

AGAIN: Suppose the stroke of a piston=36″, and that the crank has passed over 112°. How far will the piston have advanced?

The Table gives for 112° a piston position of 0.687 of the stroke.

Therefore $0.687 \times 36''=24\frac{3}{4}''$ the distance advanced by the piston while the crank has advanced 112 degrees.

There is securely fastened to the crank shaft a device called an "eccentric," which serves to impart a reciprocating motion to the slide valve. Upon close inspection it appears that this is only a mechanical subterfuge for a *small crank.*

The travel of any valve being small compared with that

of its piston, the crank required for its motion has frequently an arm or "*throw*" cb shorter than one-half the
diameter ae of the main shaft, Fig. 4.   Hence to avoid cut-

Fig. 4.

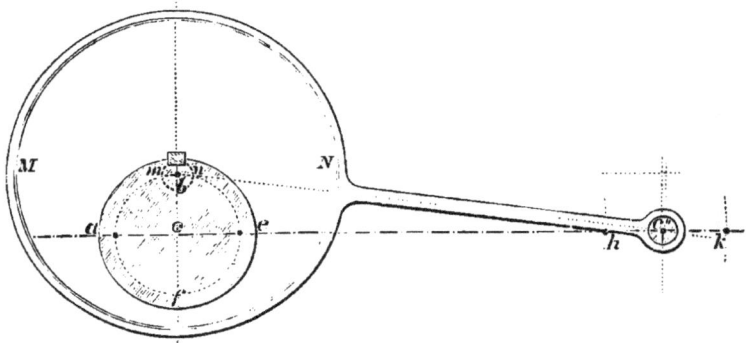

ting the shaft and the expense of forming the crank cb, the
pin m, n, and enclosing strap of the rod are greatly enlarged until they attain the common diameter M N, the
former may then be slipped on, and keyed fast to the shaft
ae.   Of course the motion will not be altered by this
change, but the same reason that led to the adoption of the
slotted cross head for tracing the piston's progress, now
compels us to substitute a small slotted cross head and rod
for the eccentric rod.   In the sequel therefore both the
crank pin and the eccentric pin (or centre of eccentric) will
be considered as transmitting their motions through slotted
cross-heads to the piston and the valve.   (See Fig. 5.)

The axes of the cylinder and of the valve stem do not
always pass through the centre of the main shaft.   When
that of the latter lies above and parallel to the former, as
shown in the figure, some expedient must be adopted for
carrying the motion of the eccentric pin up from the point
q in the central plane of the engine to e in that of the valve.

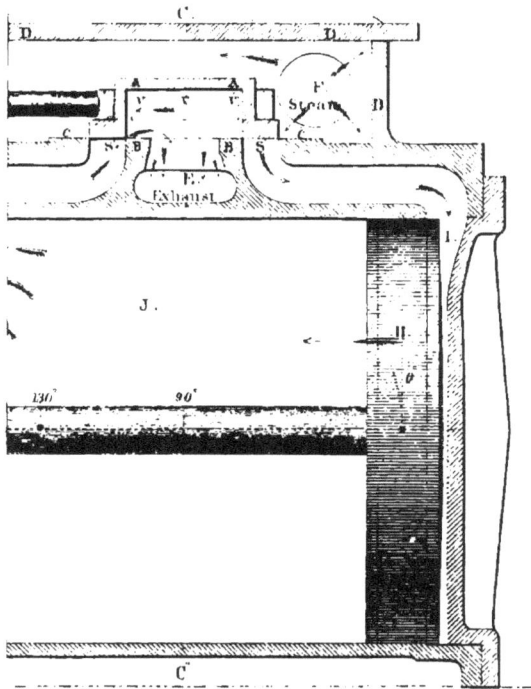

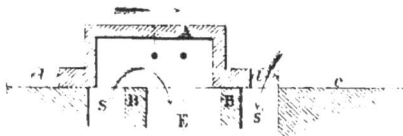

FIGURE 5

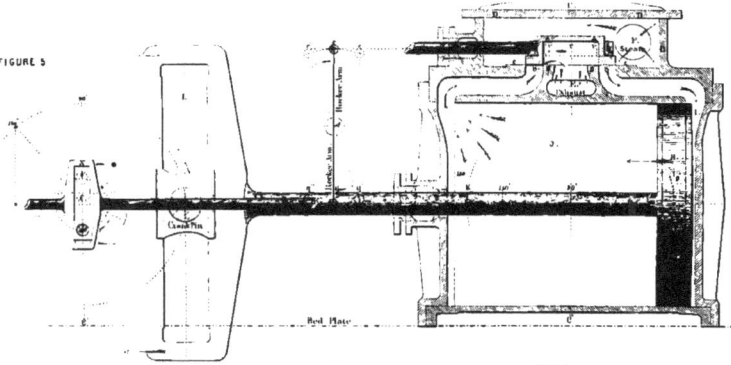

Crank Pin

Rocker Arm

Rocker Arm

Cylinder

Bed Plate

FIGURE 6.

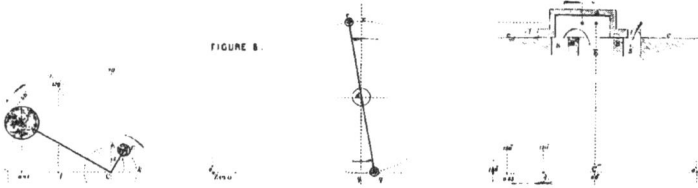

This is frequently accomplished by aid of a bar $q\,d\,e$ called a "*rocker*," free to oscillate on its firmly supported axis d. The direction of the motion then becomes the reverse of that produced by the eccentric pin and if the pins q and e are made to operate in vertical slots no irregularity will be introduced by this arrangement.

Having explained the general features of these controllers of motion, the crank and the eccentric, and having resolved them into their elementary forms, we pass to consider the parts moved and seek the law of their proportions.

The plain slide valve of a steam-engine is a device by which the entrance and exit of the steam is regulated for the opposite ends of the cylinder. It is essentially a case A, resting on a plane surface $c\,c$ as seen in cross section in Figs. 5 and 11. Through this surface are cut three passages S', S", and E, separated by the partition walls B, B, called "*bridges*." The two former lead to the opposite extremities of the cylinder, and the passage E called the "*exhaust*" leads through an oval pipe to the atmosphere. The valve A is sufficiently large to cover both the passages S', S", when standing in its neutral position. A second case D, D, called the "*steam chest*," encloses the valve A and is secured rigidly to the plane surface $c\,c$. Being larger than the valve it leaves over it much unoccupied space to which the only entrance is through the aperture F. This space is the "reception room"—so to speak—of the cylinder; to it, the steam is admitted from the boiler through F and kept in waiting during such times, as the valve in its motion completely covers the two ports.

Figure 5 represents the crank-pin at the zero point of its path, the piston at the extremity H of its stroke, the valve in the neutral position and all the parts ready for motion. A complete revolution of the crank will carry the piston

forward to K and return it to the starting point H. Whatever events take place in the journey from H to K should
be repeated in the same order on the return route from K
to H, hence in studying the motion we will seek to render
it perfect for the trip from H to K and leave the parts when
the latter point is reached in the same relative positions as
those occupied for H, so that the one will become simply a
counterpart of the other. The first point evident, is that
the port S' must be opened and again closed for the proper
admission of the steam during the stroke of the piston from
H to K; in other words, while the piston is making one
entire stroke the valve must accomplish a half and a return
half of its stroke. Such an operation can only be brought
about by securing the eccentric pin in the position f or b on
a line at right angles to the crank-arm, that of f being suitable for a direction of the crank indicated by the arrow.

Let us trace the two motions throughout one revolution
of the crank. Moving it from the zero to the 90° point will
draw the piston from the position H to the half stroke or
the line c'', c'', will advance the eccentric pin from f to k,
the rocker from $e\,q$ to $e'\,q'$, the centre of the valve from V
to V'' and completely open the port S'. As the crank progresses from 90° to 180° the eccentric pin will travel from k
towards b, gradually closing the port S' and completely
covering it when the 180° point is reached, thus leaving the
valve in the same position at the terminus of the stroke that
it occupied at the commencement. On the return stroke
from K to H the port S'' will in like manner be opened and
again closed. In thus hastily following the two entrances
of the steam to the cylinder, we have lost sight of its mode
of escape after performing the work of forcing along the
piston. Let us suppose that one revolution has been completed and the piston is prepared for a second journey from

the position H. The space J is now filled with steam and some passage of escape must be opened. This is provided in the port and pipe E, which are thrown into immediate communication with the passage S″ when the valve commences its motion, the opening becoming wider and wider as the travel progresses, only closing when the piston reaches the point K and is ready to receive fresh steam through the passage S″ for the return stroke.

Such is a brief outline of the parts and functions of the simplest form of slide valve, in which the steam is admitted at the commencement of the piston's stroke and not excluded until that stroke is completed.

This arrangement, however, is not attended with economic results, for it entirely ignores that remarkable property of steam, its elasticity. To render this latent power available, the steam should be admitted during only a portion of the piston stroke, the valve should then be closed and the confined volume of steam allowed to complete the remaining portion, by developing its power of expansion.

But how can our elementary form of valve and position of eccentric be modified for attaining this desirable result?

Suppose a cut-off were required at a piston position of 0.93 of the stroke. By carrying the crank to the 150° position (as in Fig. 6) we observe that the port S remains opened a distance l and the most ready means for effecting its closure is to lengthen the valve face by this amount. Since the cut-off must take place at relatively the same piston position in both strokes, an equal addition must be made to the other edge of the valve. Such additions to the outer edges of the valve, for the purposes of cut-off, are called overlap or simply "*lap*." The extent of this lap in the present case is evidently equal to the horizontal distance of the eccentric's centre f''' from the 90° line, because

3

without lap it would naturally close at this line. The same distance expressed in degrees would be equivalent to a "*lap angle*" of 30°.

But on referring to Fig. 6 it is clear that no such addition can be made without necessitating a change also in the eccentric location, for it would render the admission 30° *too late*. Hence if we add a lap to the valve equivalent to an eccentric motion of 30° from its neutral position, we must at the same time unkey the eccentric, and having advanced it also 30° refasten it on the main shaft. The number of degrees by which the eccentric is thus carried forward from a position at right angles to the crank-arm is termed the "*angular advance*" of the eccentric.

When the eccentric stands at right angles to the crank the exhaust closes and release commences at the *extremities* of the stroke, consequently if the eccentric be moved ahead 30° not only will the cut-off take place 30° earlier, or at a crank angle of 120° instead of 150°, but the release as well as the exhaust will take place 30° earlier or at the 150° crank angle. Although we have not secured by this process the cut-off aimed at, yet the investigation distinctly points out the means at our command for the accomplishment of *any* cut-off and will enable us to construct a Scale for determining the magnitudes of such alterations. For a cut-off of 140° there would be required an angular advance of 20° and a lap equivalent to the distance these degrees remove the eccentric centre from the line at right angles to the crank; for a cut-off of 160°, an advance of 10° with a corresponding lap, and so on; the exhaust closure taking place respectively at the 160° and 170° crank angles.

This closure of the exhaust confines the steam in the cylinder until the port is again opened for the return stroke; consequently the piston in its progress will meet with increasing resistance from the steam which it thus compresses into a less and less volume. Such opposition when properly proportioned aids in overcoming the momentum stored up in the reciprocating parts and tends to bring them economically to a state of rest at the end of each stroke. Since the closure of one port is simultaneous with the opening of the other, a release will take place of the steam which was previously impelling the piston. Within certain limits this also is conducive to a perfect action of the parts, for an early release enables a greater portion of the steam to escape before the return stroke commences, whereas a release at the end of the stroke would be attended by a resistance of the piston's progress, from the simple fact that steam *cannot* escape *instantaneously* through a small passage, but requires a certain definite portion of time dependent on the area of the opening and the pressure. The larger the opening then the less the occasion for anticipating the moment of exhaust.

We learn therefore that the moments of exhaust closure and release are, when the valve has neither " *inside lap* " nor its converse " *inside clearance*," directly dependent upon the angular advance of the eccentric, and that an angular advance of 20° produces a closure at a crank angle of 160°, one of 30° at 150° and so on, the resistance becoming continually greater as the angular advances increase. A limit at length is reached where this resistance really becomes detrimental, and an amount of power is absorbed quite inconsistent with economy of action. On this account the single eccentric is rarely used to effect cut-offs of less than $\frac{2}{3}$ the stroke. Earlier cut-offs require two valves and two eccentrics, the one set for regulating the cut-off of the steam, the other its admission and escape. This subject will be more fully discussed in Part V.

The principles just developed can be embodied in a single Diagram called the TRAVEL SCALE, whose construction is illustrated by Fig. 7.

FIG. 7.

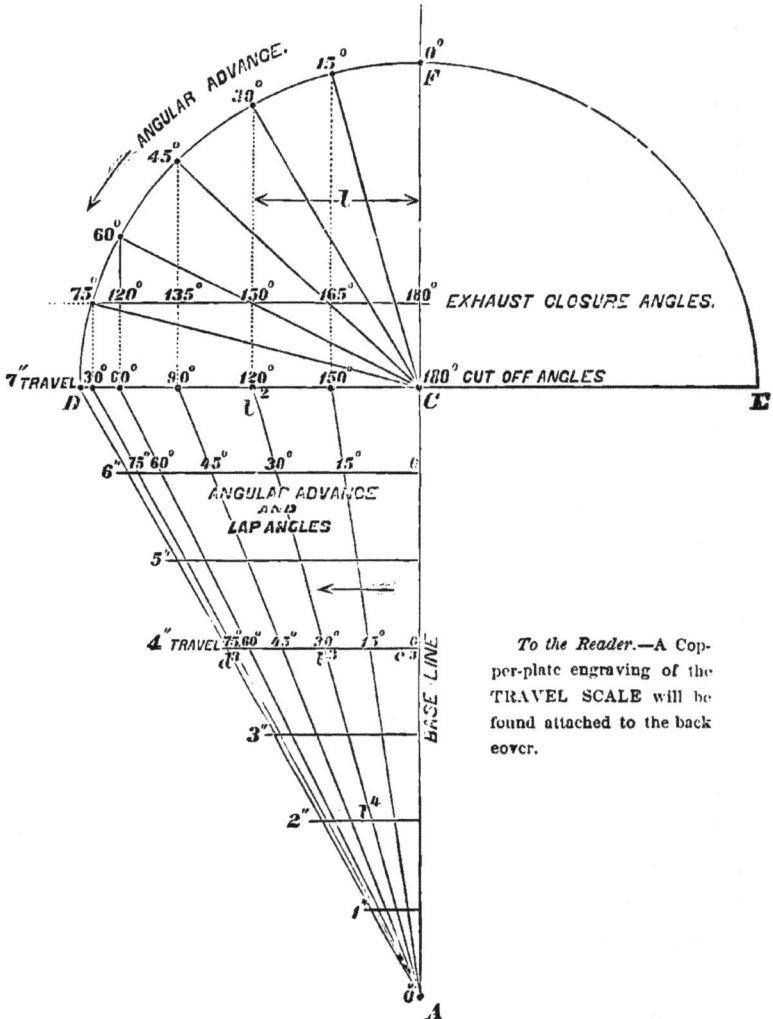

To the Reader.—A Copper-plate engraving of the TRAVEL SCALE will be found attached to the back cover.

Let E F D represent the path traversed by the centre of an eccentric whose throw equals $3\frac{1}{2}$ inches, consequently the travel of its valve=7 inches. Then C F at right angles to D E will be the normal position of the eccentric from which the angular advances must be laid off. Extend this line to some convenient point A and join the extremity D of the travel with A. Divide the line C A into 7 equal parts, and through these points draw lines parallel to D E to represent all the travels less than 7 inches. Finally project each degree of the arc D F upon the line D C and join the points thus found with the point A.

The distances from the Base Line C A, at which this group of lines intersect the travel lines, will indicate what lap should be given to accomplish various cut-offs, and their distances from the extreme travel line D A will give the width of the steam-port opening due to these travels and cut-offs. Thus for 7″ travel and a cut-off of 120° the eccentric must have an angular advance of 30° and the valve a lap equal to l' C, giving thereby a port opening l' D; while a travel of 4 inches with the same cut-off only requires a lap of l' C³ and has a port opening of l' d'. The exhaust closure of course takes place in both cases at a crank angle of 150, or piston position of 0.93 the stroke.

It will be observed that this SCALE may be applied with perfect accuracy to travels greater than 7 inches by making these lines represent their multiples; for instance, a 4-inch travel may stand for one of 8 or 12 inches; a 6-inch travel for one of 12 or 18 inches, and so on. In such cases the values of the true lap and lead will be double or thrice those given by the SCALE. Since the same principle holds for travels less than 2 inches, it is clear that the SCALE must apply to all possible dimensions.

A slip of paper and a pencil are the only paraphernalia of the Travel Scale. To illustrate its use take for—

EXAMPLE.

Extreme width of port opening must $=1\frac{1}{4}$ inches and the valve must cut off steam at 0.82 the stroke.

Required.—Angular advance of the eccentric, travel of valve, lap and point of exhaust closure.

Table A gives for a piston position of 0.82 the stroke a crank angle of 130°, for this cut-off an angular advance of 25° will be required (see line C D of the TRAVEL SCALE).

Apply the edge of a slip of paper to the Inch Scale and mark off the desired width of the port opening a, b, as in

FIG. 8.

ANGULAR ADVANCE.

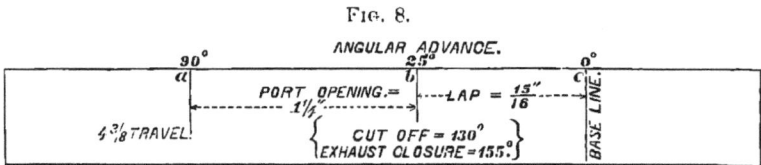

Carry the same to the Travel Scale, place the mark a over the 90° line C A and slide the edge—parallel to the line C D—until the mark b stands directly over the 25° angular advance or lap-angle line. The $4\frac{3}{8}$ inches line of travel, upon which the slip of paper here stops, will be the correct travel for the valve. Before removing the paper mark the position c of the Base line. Finally return the slip to the Inch Scale and measure the lap b c, which gives $\frac{15}{16}$ of an inch. The exhaust closure on one side and release on the other will of course take place at the 155° angle of the crank (see line C D) or at a piston position of about 0.95 of the stroke.

ANSWERS.
{
Angular Advance=25°.
Travel of valve=$4\frac{3}{8}$ inches.
Lap=$\frac{15}{16}$ inch.
Exhaust closes at 0.95 of the stroke.
}

The solution of such problems as the subjoined, will

tend to familiarize the Reader with the method of using this Travel Scale :

1st. To cut off at ⅔ the stroke, with port opening of 1½ inches.

Required.—Angular advance of the eccentric, travel of valve, lap and point of exhaust closure.

2d. To cut off at ¾ the stroke, with port opening of 1½ ins.

3d. " " ⅞ " " " "   "   3 "

4th. " " 0.7 " " " "   "   1 3⁄16 "

DIRECTION OF CRANK MOTION.

The direction of any crank motion depends on two conditions—1st. The presence or absence of a rocker for transmitting the motion ; 2d. The location of the angular advance with reference to the central line of the valve motion. Both of these may be conveniently expressed in a single Diagram like the accompanying Fig. 9, in which the posi-

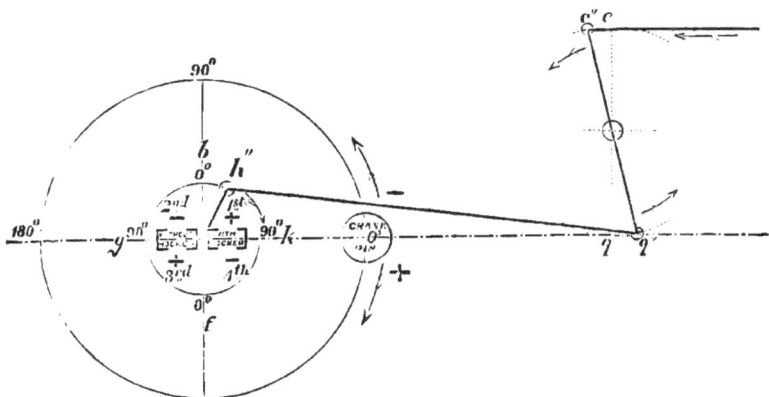

FIG. 9.

tive sign (+) represents a motion in the direction of the hands of a watch, the negative (−) a reverse motion. To produce a positive motion in any engine, whose eccentric acts through a rocker, lay off the angular advance from the line bf in the 1st quadrant (the crank standing at the zero), but for one without a rocker, the angular advance must be laid off from the same line in the 3d quadrant. The 4th and 2d quadrants in like manner belong to the negative motion. The reason for making such a disposition of the angular advance will at once appear upon tracing out either of these motions.

When the power of an engine is transmitted through a wide belt to the machinery, the direction of its crank motion will be determined by the relative locations of the main and crank shafts. The strain should *invariably* be made to fall upon the *lower* portion of the belt, the upper being thereby relaxed, sags upon its pulleys, increases the frictional surface, and materially improves the adhesion of the belt.

LEAD.

This term is applied to an alteration made in the plan of the valve motion for the purpose of concealing and neutralizing an effect, due to imperfect workmanship as well as continual wear in the boxes of the crank and cross head pins. The difficulty may be best explained with the assistance of Fig. 10.

Suppose, for instance, both boxes of the connecting rod A B, fit loosely upon the crank and cross-head pins, that the crank moving in the direction indicated by the arrow, has reached a location C A within 8 degrees of the zero, and that the piston (on account of the lost motion in the boxes) falls short of its true position B, a distance B B. If now the

momentum of the motor carries the crank-pin past its zero,
the piston, which at the moment of passage is no longer
urged or restrained by the connecting rod, will by virtue

FIG. 10.

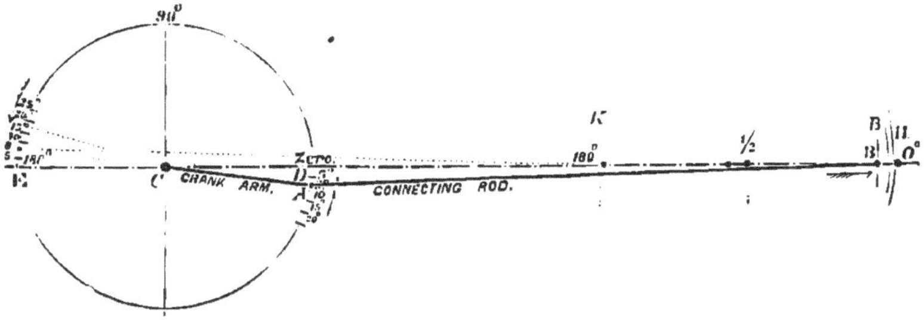

of its own momentum continue moving in the direction of
H until all the lost motion being expended, its progress is
suddenly checked and it is itself again brought under the
control of the connecting rod, which then draws it forward
upon the return stroke. These concussions are reproduced
at the end of each stroke with a degree of force and sound
directly dependent on the extent of the lost motion and the
momentum of the piston with its connecting rod. Where
the parts are of great weight, as in a marine engine, the
sound becomes very loud and the engine is said to
"*thump*" or "*pound*" on the "*centres.*" Two ways present themselves for counteracting this effect : the one, by
making the boxes so durable and the workmanship so perfect that lost motion becomes almost impossible ; the other,
by introducing a resistance to the momentum of the piston
capable of *completely* overcoming it before the end of the
stroke, in other words by allowing the steam to enter the
cylinder a short time *previous* to the termination of the
stroke. With small engines the first method is practicable,
but in large ones both are more commonly employed be-

cause with these, a very small amount of lost motion suffices to produce a disagreeable sound.

The width of port opening given by any valve at the moment its crank passes either centre, is called the "*lead*" of the valve ; and the angular distance of the crank from its zero at the instant this opening commences, the "*lead angle.*"

The opening together with the angle (or time) limit the power of the steam in its effect upon the lost motion ; for even a small opening continued through a long time may prove as efficient for the admission as a large opening during a very short time.

Since sound, the effect of lost motion, depends upon the weight and velocity of the reciprocating parts, the lead requisite must vary for different engines and also for the same engines at different velocities. The *exact* amount *cannot* be predicated in any particular case, but after the engine has been constructed it may be experimentally determined by gradually increasing the angular advance of the eccentric until some position is found which results in a smooth and noiseless movement of the reciprocating parts. We have before alluded to the effect of compression by a premature closure of the exhaust, but it must be distinctly understood that this agency unassisted cannot neutralize the evils of lost motion without injuring the admission of the return stroke. In this respect it differs from lead. It should then usually be supplemented by lead in order to accomplish a smooth action of the parts and free opening of the steam port for the return stroke. Observe also, that so long as the lead angle amounts to only a few degrees no impression can be produced on the continuity of the crank motion, for the lever arm will be too small for the power to exert any influence over the crank.

The limits of the *lead angle* are commonly *zero and 8 for stationary engines;* while for any given angle the width of opening will depend upon the travel of the valve and the point of its cut-off.

It remains to be shown that the Travel Scale is quite as applicable to valves having a certain lead as to those without any. Referring again to Fig. 6, imagine an increase in its angular advance of 5°, the valve will then close at 115° instead of 120° and reopen its port 5° *before* the crank reaches the extremity of the stroke; but if the lap be reduced 5° when the angular advance is increased 5 , the cut off will still remain 120°, while the port commences to open 10° *before* the end of the stroke. Consequently if we wish to arrange a valve for a certain number of degrees lead, without altering the point of cut-off, it will simply be necessary to *find the angular advance for a valve without lead, add ½ the lead angle for a new angular advance, and subtract the other ½ for an angle by which to measure the lap.*

If in the Example of Fig. 8 a *lead of 8 degrees* had been required, with the *same* cut-off, the angular advance

$$\left. \begin{array}{l} \text{would have become } 25°+4°=29° \\ \text{and the lap angle} \quad 25°-4°=21° \end{array} \right\}$$

and by applying the port opening marks *a* and *b* to the 90° and 21° lines,—instead of the 90° and 25° lines,—we would have obtained a travel of $3\frac{1}{4}$ inches and a lap of $\frac{11}{16}$ inch; while the distance between the angular advance line of 29° and the lap angle line of 21° would have equaled ¼ inch, the *width* of the *lead opening* at the extremity of the piston stroke.

The change in the angular advance of course changes the exhaust closure from 155° to 151° or about 0.93 of the piston's stroke.

Supposing then a lead angle of 8° for the same problem the answers become :—

Angular Advance........................... = 29°.
Lap angle........   .... = 21 .
Travel................................... = 5.7 inches.
Lap..................................... = 11/16 inch.
Lead.................................... = 1/4 inch.
  Exhaust closes at 0.93 of the stroke.

———————

Similar suppositions made and applied to the other trial problems will give all the practice requisite for successfully using the TRAVEL SCALE.

It seems almost unnecessary to observe that the SCALE effects with equal readiness and precision solutions directly the converse of that just accomplished. Thus, if the above lap, lead and travel were given, to determine the exhaust closure and cut-off, we would mark the lap and lead on a strip of paper as in Fig. 12, apply the same to the 3⅝ inch travel line of the scale, which would show at once an angular advance of 29° and consequently exhaust closure of 0.93 the stroke ; also a lap angle of 21° with lead. or 25° without, the same as a cut-off of 130°=0.82 the entire stroke.

A moment's reflection will also show that—during the progress of the crank—the varying width of the port opening from the simple lead out to the maximum width and back again to the period of cut-off, might *readily* be traced on the SCALE, and all the information common to the popular method of ellipse or other construction, be immediately obtained. But the facts thus gained, would prove of very *trifling* moment, so long as the valve had received a correct maximum port opening.

WIDTH OF BRIDGE.

This dimension is usually made of equal thickness with the cylinder, in order to secure a perfect casting, but at times it becomes necessary to increase its width. The only danger from a narrow bridge is an *overtravel* of the valve, by which the exhaust passage would be placed in direct communication with the "live steam" in the chest, and followed by continual waste of the power. Obviously this cannot occur while the difference between the port opening and the steam port does not exceed the width of the bridge. (Fig. 11.) But to prevent even the possibility of a leakage :—

Add about ¼ of an inch to the width of the opening and from their sum subtract the width of the steam port.

Thus the width of the steam port in the example of Fig. 5. should have been at least :—

$$1\tfrac{1}{4} + \tfrac{1}{4}'' - 1'' = \tfrac{1}{2} \text{ inch.}$$

When however the width of the opening is less than that of the steam port, the danger of such an escape entirely vanishes.

———

WIDTH OF EXHAUST PORT.

The main difficulty to be avoided in proportioning the width of this port is the possibility of a reduction in its area, when the valve attains extreme travel, to an opening materially less than that of the steam port from which it derives its supply.

Suppose that the valve in Figure 11 has reached the end of its half travel, or the exhaust edge V moved a distance

R from its neutral position V^2; then by the above condition, E will evidently equal $(S+R-B)$.

Which furnishes the following general

RULE

For determining width of Exhaust port.

Add the width of the steam port to ½ the travel and from their sum subtract the width of the bridge.

When called upon to perform the addition or subtraction of many fractional portions of an inch, it will generally be found more convenient to express these decimally than by those *very awkward* subdivisions sixty-fourths, thirty-seconds, etc.

Fractions of an inch expressed decimally.

$\frac{1}{64}$ of inch	=.0156	$\frac{1}{4}+\frac{3}{32}$ of inch	=.3438	$\frac{5}{8}+\frac{1}{16}$ of inch	= .6875
$\frac{1}{32}$ "	=.0313	$\frac{3}{8}$ "	=.375	$\frac{5}{8}+\frac{3}{32}$ "	= .7188
$\frac{1}{16}$ "	=.0625	$\frac{3}{8}+\frac{1}{32}$ "	=.4063	$\frac{3}{4}$ "	= .75
$\frac{3}{32}$ "	=.0938	$\frac{7}{16}$ "	=.4375	$\frac{3}{4}+\frac{1}{32}$ "	= .7813
$\frac{1}{8}$ "	=.125	$\frac{7}{16}+\frac{3}{32}$ "	=.4688	$\frac{3}{4}+\frac{1}{16}$ "	= .8125
$\frac{1}{8}+\frac{1}{32}$ "	=.1563	$\frac{1}{2}$ "	=.5	$\frac{3}{4}+\frac{3}{32}$ "	= .8438
$\frac{1}{8}+\frac{1}{16}$ "	=.1875	$\frac{1}{2}+\frac{1}{32}$ "	=.5313	$\frac{7}{8}$ "	= .875
$\frac{1}{8}+\frac{3}{32}$ "	=.2188	$\frac{1}{2}+\frac{1}{16}$ "	=.5625	$\frac{7}{8}+\frac{1}{32}$ "	= .9063
$\frac{1}{4}$ "	=.25	$\frac{1}{2}+\frac{3}{32}$ "	=.5938	$\frac{7}{8}+\frac{1}{16}$ "	= .9375
$\frac{1}{4}+\frac{1}{32}$ "	=.2813	$\frac{5}{8}$ "	=.625	$\frac{7}{8}+\frac{3}{32}$ "	= .9688
$\frac{1}{4}+\frac{1}{16}$ "	=.3125	$\frac{5}{8}+\frac{1}{32}$ "	=.6563	1 *inch*	= 1.000

INSIDE LAP.

The effect on a valve motion of inside lap is to—

Prolong the Expansion, and

Hasten the Compression.

(A contrary effect for inside clearance.)

The former is occasionally added in the case of high-speed engines having very late cut-offs. In such instances the compression is arranged to commence at about ⅚ of the stroke, or at an angle of 138 degrees, and the release at an angle not exceeding 160°. For example, if the angular advance equals 32° (with a travel of 4⅝ inches) the compression would commence at a crank angle of 148° or 10° later than the above limit: hence if we give the valve an inside lap of 10° or ⅔ of an inch found as in Fig. 12, the expan-

FIG. 12.

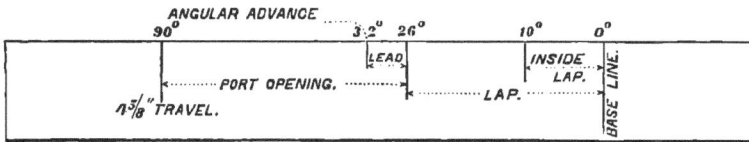

sion will continue from the point of cut-off to 148 +10 = 158 degrees, and the compression commence at 148° −10°= 138 degrees, instead of both events taking place at the 148° angle of the crank.

We think the foregoing investigations fully sustain our remarking in conclusion that any questions, relating to the travel of the valve, the varying widths of the exhaust and steam-port openings for every possible position of the crank, the moments of closure and release, and other points of interest, can not only be determined with *perfect* precision by means of the TRAVEL SCALE, but their solution will prove well nigh *instantaneous* when compared with the indirect and tedious methods that have heretofore obtained in popular usage.

4

GENERAL EXAMPLE.

What dimensions should be given to the cylinder and valve of an engine like Fig. 5 to secure an indicated horse power of 150 with

Pressure of steam in boiler at 65 lbs. ;

The crank to make 50 revolutions per minute, and the steam to be cut off at $\frac{2}{3}$ the stroke?

The mean effective pressure (page 16) $=65 \times 0.82 = 53.3$ lbs. Piston speed (page 17) $=$ say 250 ft. per minute. Area of piston, page 18,

$$A = \frac{33,000 \times 150}{250 \times 53.3} = \frac{132 \times 150}{53.3} = 371 \text{ sq. inches.}$$

Therefore diameter of piston $=21\frac{3}{4}$ ins., say 22 inches.

Stroke of piston (page 19) $=\frac{250}{100} = 2.5$ ft. $= 2$ ft. 6 inches.

Port area (page 22) $=371$ sq. inches $\times .047 = 17.4$ sq. inches.

If the length of the steam port $=20$ inches then its width will $=\frac{17.4}{20} = \frac{7}{8}$ inch.

Width of port opening W (by page 23) may vary between 0.6 and 0.9 the width of the entire port, but for the sake of greater precision in the cut-off and freer opening of the port at the commencement of the stroke, let us make its width equal about 1.5 width of steam port, or—

$$W = 1.5 \times \tfrac{7}{8}'' = 1\tfrac{1}{4} \text{ inches.}$$

Area of steam pipe (page 22) $=371$ sq. inches $\times .032 = 11.9$ square inches.

Area of exhaust pipe $=$ area of steam port $=17.4$ sq. ins.

The respective diameters of these pipes will therefore be 4 and $4\frac{3}{4}$ inches. By the Travel Scale, the angular advance for the given cut-off of 110° equals (without lead) 35° and with a lead angle of say 6°,

Angular advance will$=35°+3°=38$ degrees,
And lap angle will$=35°-3°=32$ degrees.

Now apply the width of port opening $1\frac{1}{4}$ inches to the 90° and 32 lines of the Travel Scale, as page 38, and we find that the Travel must$=5\frac{3}{4}$ inches.

After marking the Base line and angular advance we have—

Lap$=1\frac{7}{16}$ inches : lead$=\frac{1}{4}$ inch.

The bridge, page 47, should not be less than $\frac{1}{4}''+1\frac{1}{4}''-\frac{7}{8}''$ $=\frac{5}{8}$ inch. If, however, the cylinder has a thickness of 1 inch the bridge must be made of the same width.

Width of exhaust port, page 48,

$$E=\tfrac{5}{8}+2\tfrac{3}{4}-1''=2\tfrac{5}{8} \text{ inches.}$$

Also we have the width of each valve face F and N$=$width of steam port$+$lap)

$$\text{Equals, } \tfrac{5}{8}''+1\tfrac{7}{16}''=2\tfrac{5}{16} \text{ inches,}$$

And the total length of the valve or

L$=$exhaust port $+$ 2 bridges $+$ 2 faces$=2\frac{5}{8}+2+4\frac{5}{8}=9\frac{1}{4}$ inches.

The angular advance being 38° the exhaust will close and release commence at the 142° angle of the crank (see Travel Scale) or at 0.895 of the stroke$=30''\times0.895=26\frac{7}{8}$ inches and the cut-off take place at $\frac{2}{3}\times30''=20$ inches; which embraces all the required dimensions.

PART II.

SHORT-HAND METHOD

FOR

VALVE PROPORTIONS.

SHORT-HAND METHOD.

THE following table has been prepared by means of the Travel Scale: and embodies all its essential features.

For a cut-off	Valve travel, should be:		Lap, should be:		The Exhaust will close at:
0.5 = ½ stroke.	6.6 times	{ width of port opening.	2.3 times	{ width of port opening.	0.85 stroke.
0.55 "	6 "	"	2 "	"	0.87 "
0 625= ⅝ "	5.3 "	"	1.6 "	"	0.89 "
0.64 "	5 "	"	1.5 "	"	0.9 "
0.666= ⅔ "	4.7 "	"	1.35 "	"	0.91 "
0.7 "	4.4 "	"	1.2 "	"	0.92 "
0.75 = ¾ "	4 "	"	1 "	"	0.93 "
0.8 "	3 6 "	"	0.82 "	"	0.94 "
0.83 "	3.4 "	"	0 7 "	"	0.95 "
0.875= ⅞ "	3.1 "	"	0.54 "	"	0.96 "
0.9 "	3 "	"	0.45 "	"	0.97 "

It will be remembered that in dealing with crank and piston motions we regarded the stroke as equal to *Unity* and their positions (at certain important periods) as decimal portions of the entire stroke. In this chapter we have set aside all consideration of lead, and made the extreme opening of the steam port by the valve, a *Unit* for measuring, how much travel and lap are necessary for a given cut-off?

Take for example any extreme port opening—say $1\frac{1}{2}$ inches for a cut-off at $\frac{3}{4}$ stroke?

We see by a glance at the table that the travel must be 4 times as great as the port opening, while the lap must be *once* the port opening, thus giving instantly the

Answers : $\begin{cases} \text{Travel} = 6 \text{ inches.} \quad \text{Lap} = 1\frac{1}{2} \text{ inches.} \\ \text{Exhaust closes at 0.93 stroke.} \end{cases}$

EXAMPLES FOR PRACTICE.

Port opening $= 1$ inch, Cut-off $= \frac{1}{2}$ stroke—Find the Travel, and lap.?

" $= 1\frac{1}{2}$ " " $= \frac{5}{8}$ " " " " " ?

" $= 2$ " " $= \frac{7}{8}$ " " " " " ?

SHIFTING ECCENTRIC FOR PORTABLE ENGINES.

FIG. B.

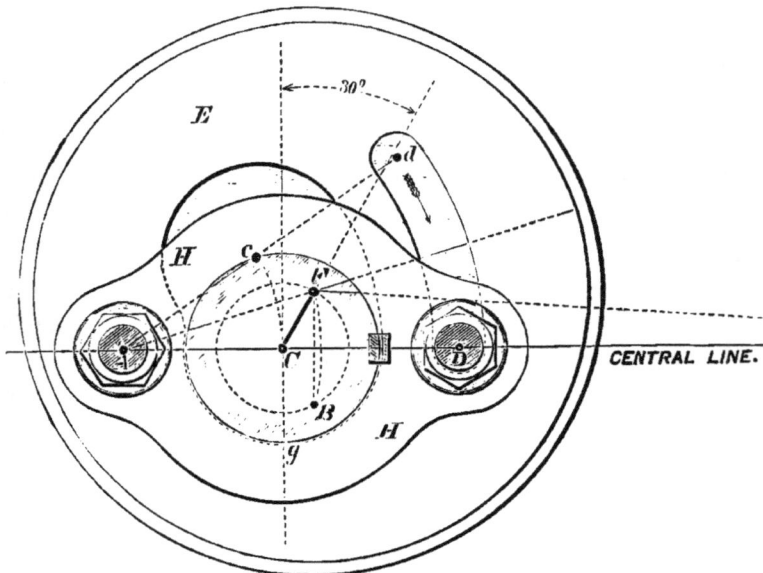

CENTRAL LINE.

PART III.

GENERAL PROPORTIONS

MODIFIED BY

CRANK AND PISTON CONNECTION.

THUS far we have confined our attention to a form of connection called the "slotted cross-head," and have been able therewith to deduce laws governing the proportions of the various parts of the valve, as well as to devise a most simple and rapid method for determining their magnitudes. But since this connection seldom obtains in practice, it becomes necessary for us to analyze the form shown in Fig. 13, to modify their general proportions to accord with the new conditions and to eliminate as far as possible all the irregularities they tend to create.

It will be observed, by inspecting this Figure, that the cross head pin is drawn a distance BB″ beyond its half stroke position B″, when the crank attains an angle of 90°, that this irregularity is due to the want of parallelism of the connecting rod, with its original position—during the progress of the crank pin in its semi-revolution—and that a rod of virtually infinite length produces a motion of the piston identical with that of the cross-head. It follows that the irregularity BB″ will vary with the different ratios that may exist between the length of the crank arm and the connecting rod. In subsequent comparisons of these two terms, the length of the *crank arm* will always be regarded

as the *Unit measure* and that of the connecting rod as a
certain number of times the length of the crank arm.

Let the crank arm C A be equal to unity and the connecting rod A B=4, then their ratio is that of 1 to 4, (1 : 4.)
When the arm occupies the 90° position the cross-head pin
will be drawn a distance B B″ beyond the half stroke point

FIG. 13.

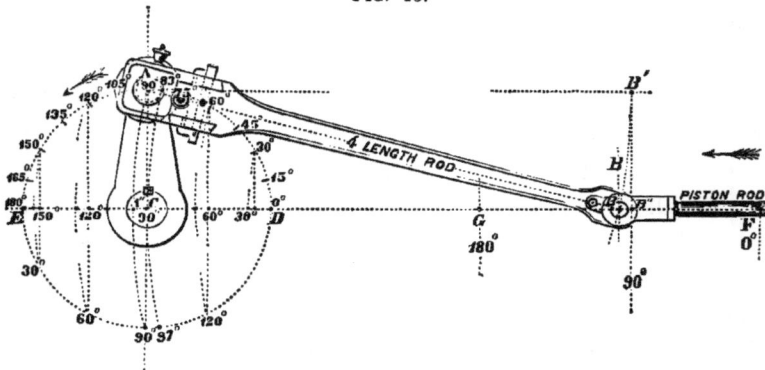

B″. With B as a centre and A B as radius, describe the
arc A A″. If the occasion required, it might be readily
proved that A″, the point of its intersection with the line
D E, is the same distance from C that B is from B″. Placing
the crank in other positions—as at 30°, 60°, 120° and so on
—and describing similar arcs there will result like irregularities but of a less degree, all of which however vanish
at the extremities of the stroke D and E. It becomes evident
therefore that the effect of this form of connection is ; *to
carry the piston ahead of its proper positions throughout
the forward stroke and on the return stroke to make it lag
behind the positions due to the locations of the crank pin.*

Consequently the *one crank angle*, for a given piston
position (as in Table A), will no longer serve both the
forward and return strokes, but a *new* table must be

constructed which shall furnish at sight the proper angles of the crank for various piston positions in both the Forward and the Return strokes, and these for every important ratio of crank to connecting rod between 1 : 4 and 1 : 8 with which intermediate values may readily be determined by interpolation. Such is presented in the following STROKE TABLE.

The fractional portions of a degree have been given as small as can conveniently be laid off with a protractor.

By transposing the terms *Forward* and *Return* the angles in the Table will apply to the case of a " *Back Action* " *Engine.* For the irregularities of the motion are necessarily *reversed* in such instances, because the cross head and cylinder lie on *opposite* sides of the main shaft instead of on the *same* side.

FIRST EXAMPLE.

The connecting rod of a certain engine=8' 3"=99".
The crank arm=18 inches.
Cut-off takes place at 0.65 of the stroke.

Required—The forward and return stroke crank angles.

Divide length of connecting rod by that of the crank arm : thus

$$\frac{99}{18} = 5\frac{1}{2}$$

Their ratio therefore will be that of 1 : $5\frac{1}{2}$.

STROKE TABLE.

Piston Position. (Stroke = unity.)	CRANK ANGLES. (FOR ORDINARY CONNECTING ROD.)								
	RATIO 1 : 4.			RATIO 1 : 4½.			RATIO 1 : 5.		
	Forward	Return.	Diff.	Forward	Return.	Diff.	Forward	Return.	Diff.
	deg.	deg.	deg.	deg.	deg.	deg.	deg.	deg.	deg.
$0.125 = \frac{1}{8}$	$37\frac{3}{4}$	$46\frac{3}{4}$	$9\frac{3}{4}$	$37\frac{5}{8}$	$46\frac{1}{4}$	$8\frac{5}{8}$	$37\frac{7}{8}$	$45\frac{5}{8}$	$7\frac{1}{4}$
0.2	48	$59\frac{1}{4}$	$11\frac{1}{4}$	$48\frac{1}{4}$	$58\frac{3}{4}$	$10\frac{1}{4}$	$48\frac{3}{4}$	$58\frac{1}{4}$	$9\frac{3}{4}$
$0.25 = \frac{1}{4}$	$54\frac{3}{4}$	$66\frac{3}{4}$	$12\frac{3}{4}$	$54\frac{3}{4}$	66	$11\frac{7}{8}$	$55\frac{3}{4}$	$65\frac{3}{8}$	10
0.3	$60\frac{3}{4}$	$73\frac{3}{4}$	13	61	$72\frac{5}{8}$	$11\frac{5}{8}$	$61\frac{1}{4}$	72	$10\frac{1}{4}$
$0.333 = \frac{1}{3}$	$64\frac{1}{4}$	$77\frac{5}{8}$	$13\frac{3}{8}$	$64\frac{3}{4}$	$76\frac{7}{8}$	$12\frac{1}{4}$	$65\frac{3}{4}$	$76\frac{1}{4}$	$10\frac{7}{8}$
$0.375 = \frac{3}{8}$	$68\frac{7}{8}$	$82\frac{3}{8}$	$13\frac{1}{2}$	$69\frac{1}{4}$	82	$12\frac{1}{2}$	$70\frac{1}{4}$	$81\frac{3}{8}$	$11\frac{1}{4}$
0.4	$71\frac{1}{4}$	$85\frac{1}{2}$	$13\frac{1}{2}$	$72\frac{3}{4}$	$84\frac{7}{8}$	$12\frac{1}{4}$	73	$84\frac{1}{4}$	$11\frac{1}{4}$
0.45	$77\frac{3}{4}$	$91\frac{1}{2}$	$14\frac{1}{4}$	$78\frac{1}{4}$	$90\frac{5}{8}$	$12\frac{1}{4}$	$78\frac{5}{8}$	90	$11\frac{3}{4}$
$0.5 = \frac{1}{2}$	$82\frac{7}{8}$	$97\frac{1}{4}$	$14\frac{1}{4}$	$83\frac{3}{4}$	$96\frac{1}{4}$	$12\frac{1}{4}$	$84\frac{3}{4}$	$95\frac{5}{8}$	$11\frac{3}{4}$
0.55	$88\frac{1}{4}$	$102\frac{3}{8}$	$14\frac{1}{4}$	$89\frac{3}{4}$	$101\frac{7}{8}$	$12\frac{1}{4}$	90	$101\frac{7}{8}$	$11\frac{3}{4}$
0.6	$94\frac{3}{4}$	$108\frac{1}{4}$	$13\frac{7}{8}$	$95\frac{1}{4}$	$107\frac{5}{8}$	$12\frac{1}{4}$	$95\frac{3}{4}$	107	$11\frac{1}{4}$
$0.625 = \frac{5}{8}$	$97\frac{1}{4}$	$111\frac{1}{4}$	$13\frac{7}{8}$	98	$110\frac{1}{4}$	$12\frac{1}{4}$	$98\frac{5}{8}$	$109\frac{3}{4}$	$11\frac{1}{4}$
0.65	$100\frac{1}{4}$	$113\frac{7}{8}$	$13\frac{5}{8}$	$101\frac{1}{4}$	113	$12\frac{1}{4}$	$101\frac{3}{4}$	$112\frac{1}{2}$	$10\frac{7}{8}$
$0.666 = \frac{2}{3}$	$102\frac{3}{4}$	$115\frac{1}{4}$	$13\frac{3}{8}$	$103\frac{1}{4}$	$115\frac{1}{4}$	$12\frac{2}{5}$	$103\frac{3}{4}$	$114\frac{1}{2}$	$10\frac{7}{8}$
0.68	104	$117\frac{3}{4}$	$13\frac{3}{8}$	$104\frac{3}{4}$	$116\frac{3}{4}$	12	$105\frac{1}{2}$	$116\frac{1}{4}$	$10\frac{3}{4}$
0.7	$106\frac{1}{2}$	$119\frac{1}{2}$	13	$107\frac{3}{8}$	119	$11\frac{5}{8}$	108	$118\frac{1}{4}$	$10\frac{1}{4}$
0.71	$107\frac{7}{8}$	$120\frac{3}{4}$	$12\frac{7}{8}$	$108\frac{5}{8}$	$120\frac{1}{4}$	$11\frac{5}{8}$	$109\frac{3}{4}$	$119\frac{3}{4}$	$10\frac{3}{8}$
0.73	$110\frac{1}{4}$	$123\frac{1}{4}$	$12\frac{3}{4}$	$111\frac{1}{4}$	$122\frac{5}{8}$	$11\frac{3}{4}$	112	$122\frac{1}{4}$	$10\frac{1}{4}$
$0.75 = \frac{3}{4}$	$113\frac{1}{4}$	$125\frac{5}{8}$	$12\frac{3}{8}$	114	$125\frac{1}{4}$	$11\frac{1}{4}$	$114\frac{5}{8}$	$124\frac{1}{4}$	10
0.76	$114\frac{5}{8}$	$126\frac{3}{4}$	$12\frac{1}{4}$	$115\frac{3}{4}$	$126\frac{3}{4}$	11	$116\frac{1}{4}$	$125\frac{7}{8}$	$9\frac{7}{8}$
0.77	$116\frac{1}{4}$	$128\frac{1}{4}$	12	$116\frac{7}{8}$	$127\frac{5}{8}$	$10\frac{3}{4}$	$117\frac{1}{4}$	$127\frac{1}{4}$	$9\frac{3}{4}$
0.78	$117\frac{5}{8}$	$129\frac{3}{8}$	$11\frac{3}{4}$	$118\frac{3}{4}$	$128\frac{7}{8}$	$10\frac{1}{2}$	119	$128\frac{1}{4}$	$9\frac{1}{2}$
0.79	$119\frac{1}{4}$	$130\frac{1}{4}$	$11\frac{1}{2}$	$119\frac{7}{8}$	$130\frac{1}{4}$	$10\frac{3}{8}$	$120\frac{1}{4}$	$129\frac{3}{4}$	$9\frac{1}{4}$
0.8	$120\frac{1}{4}$	132	$11\frac{1}{4}$	$121\frac{1}{4}$	$131\frac{5}{8}$	$10\frac{1}{4}$	$121\frac{7}{8}$	$131\frac{1}{4}$	$9\frac{1}{4}$
0.81	$122\frac{1}{4}$	$133\frac{1}{4}$	$11\frac{1}{8}$	$122\frac{7}{8}$	$132\frac{1}{2}$	10	$123\frac{1}{4}$	$132\frac{1}{4}$	9
0.82	$123\frac{5}{8}$	$134\frac{3}{8}$	11	$124\frac{1}{4}$	$134\frac{1}{4}$	$9\frac{3}{4}$	125	$133\frac{3}{4}$	$8\frac{3}{4}$
0.83	$125\frac{3}{4}$	136	$10\frac{5}{8}$	126	$135\frac{5}{8}$	$9\frac{5}{8}$	$126\frac{5}{8}$	$135\frac{1}{4}$	$8\frac{3}{4}$
0.84	127	$137\frac{1}{4}$	$10\frac{1}{4}$	$127\frac{5}{8}$	137	$9\frac{3}{4}$	$128\frac{1}{4}$	$136\frac{3}{4}$	$8\frac{1}{4}$
0.85	$128\frac{3}{4}$	$138\frac{7}{8}$	$10\frac{1}{4}$	$129\frac{3}{4}$	$138\frac{1}{4}$	$9\frac{1}{4}$	130	$138\frac{1}{4}$	$8\frac{1}{4}$
0.86	$130\frac{1}{4}$	$140\frac{3}{4}$	$9\frac{7}{8}$	$131\frac{1}{4}$	140	$8\frac{3}{4}$	$131\frac{5}{8}$	$139\frac{3}{4}$	$8\frac{1}{4}$
0.87	$132\frac{1}{4}$	142	$9\frac{5}{8}$	133	$141\frac{3}{4}$	$8\frac{3}{4}$	$133\frac{1}{2}$	141	$7\frac{3}{4}$
$0.875 = \frac{7}{8}$	$133\frac{1}{4}$	$142\frac{5}{8}$	$9\frac{3}{4}$	$133\frac{3}{4}$	$142\frac{3}{4}$	$8\frac{5}{8}$	$134\frac{3}{4}$	142	$7\frac{1}{4}$
0.88	$134\frac{1}{4}$	$143\frac{1}{2}$	$9\frac{1}{4}$	$134\frac{1}{4}$	$143\frac{1}{4}$	$8\frac{1}{4}$	$135\frac{1}{4}$	$142\frac{7}{8}$	$7\frac{5}{8}$
0.89	$136\frac{1}{4}$	$145\frac{1}{4}$	9	$136\frac{3}{4}$	$144\frac{3}{4}$	8	$137\frac{1}{4}$	$144\frac{1}{2}$	$7\frac{1}{4}$
0.9	$138\frac{1}{4}$	$146\frac{3}{4}$	$8\frac{5}{8}$	$138\frac{5}{8}$	$146\frac{3}{4}$	$7\frac{3}{4}$	$139\frac{1}{4}$	$146\frac{1}{4}$	7
0.91	$140\frac{1}{4}$	$148\frac{1}{2}$	8	141	$148\frac{1}{4}$	$7\frac{1}{4}$	$141\frac{5}{8}$	148	$6\frac{5}{8}$
0.92	$142\frac{1}{4}$	$150\frac{3}{4}$	$7\frac{3}{4}$	$143\frac{1}{4}$	$150\frac{1}{4}$	$6\frac{7}{8}$	$143\frac{1}{4}$	$149\frac{7}{8}$	$6\frac{1}{4}$
0.95	$150\frac{3}{8}$	$156\frac{3}{4}$	$6\frac{1}{8}$	$151\frac{1}{8}$	$156\frac{1}{2}$	$5\frac{3}{4}$	$151\frac{1}{4}$	$156\frac{3}{8}$	$4\frac{7}{8}$

STROKE TABLE.

Piston Position. (Stroke = unity.)	Crank Angles. (For ordinary connecting rod.) Ratio 1 : 5¼. Forward	Ratio 1 : 5¼. Return.	Ratio 1 : 5¼. Diff.	Ratio 1 : 6. Forward	Ratio 1 : 6. Return.	Ratio 1 : 6. Diff.	Ratio 1 : 6½. Forward	Ratio 1 : 6½. Return.	Ratio 1 : 6½. Diff.
	deg.	deg.	deg.	deg.	deg.	deg.	deg.	deg.	deg.
0.125 = ⅛	38¼	45¼	7	38½	44¾	6¼	38¾	44½	5¾
0.2	49¼	57½	8¼	49½	57¼	7¾	49¾	56¾	7
0.25 = ¼	55¾	64¾	9	56¼	64⅜	8⅛	56⅜	64	7⅝
0.3	61⅞	71½	9⅝	62⅜	71	8⅝	62⅝	70⅝	8
0.333 = ⅓	65⅞	75⅝	9¾	66¼	75¼	9	66⅝	74¾	8⅛
0.375 = ⅜	70⅜	80¾	10⅜	71⅛	80¼	9⅛	71⅜	79⅝	8¼
0.4	73½	83⅝	10⅛	73⅝	83¼	9⅝	74¼	82¾	8½
0.45	79¼	89½	10¼	79⅜	88⅞	9½	80	88⅝	8⅝
0.5 = ½	84⅞	95⅛	10¼	85⅜	94⅝	9¼	85⅝	94⅜	8¾
0.55	90½	100¾	10¼	91⅛	100⅜	9¼	91⅜	100	8⅝
0.6	96⅜	106½	10⅛	96¾	106⅛	9⅜	97¼	105¾	8½
0.625 = ⅝	99¼	109⅜	10⅛	99¾	108⅞	9⅛	100⅛	108⅝	8½
0.65	102⅜	112¼	9⅞	102¾	111¾	9	103⅛	111½	8⅜
0.666 = ⅔	104⅜	114⅛	9¾	104¾	113¾	9	105¼	113⅜	8⅛
0.68	106⅛	115¾	9⅝	106½	115⅜	8⅞	106⅞	115	8⅛
0.7	108½	118⅛	9⅝	109	117⅝	8⅝	109⅜	117⅜	8
0.71	109⅞	119¼	9⅜	110⅜	118⅞	8½	110¾	118⅝	7⅞
0.73	112½	121¾	9¼	113	121⅜	8⅜	113⅜	121⅛	7¾
0.75 = ¾	115¼	124¼	9	115⅝	123¾	8⅛	116	123⅝	7⅝
0.76	116⅝	125½	8⅞	117	125⅛	8⅛	117⅜	124⅞	7½
0.77	118	126¾	8¾	118½	126½	8	118⅞	126¼	7⅜
0.78	119½	128⅛	8⅝	120	127¾	7¾	120⅜	127½	7⅛
0.79	121	129⅜	8⅜	121⅜	129	7⅝	121¾	128¾	7
0.8	122½	130¾	8¼	122⅞	130½	7⅝	123¼	130¼	7
0.81	124	132	8	124⅜	131¾	7⅜	124⅝	131½	6⅞
0.82	125½	133½	8	126	133⅛	7⅛	126¼	132⅞	6⅝
0.83	127⅛	134¾	7⅝	127½	134½	7	127¾	134⅜	6⅝
0.84	128¾	136⅜	7⅝	129¼	136	6¾	129½	135¾	6¼
0.85	130⅜	137¾	7⅜	130¾	137½	6½	131¼	137⅜	6⅛
0.86	132¼	139⅜	7⅛	132½	139	6½	132¾	138⅞	6⅛
0.87	133⅞	141	7⅛	134¼	140⅝	6⅜	134½	140½	6
0.875 = ⅞	134¾	141¾	7	135¼	141½	6¼	135½	141¼	5¾
0.88	135¾	142⅝	6⅞	136¼	142⅜	6⅛	136½	142¼	5¾
0.89	137⅝	144¼	6⅝	138	144	6	138¼	143¾	5½
0.9	139¾	145⅞	6⅛	140⅛	145⅝	5½	140⅜	145½	5⅛
0.91	141¾	147¾	6	142⅛	147½	5⅜	142⅜	147⅜	5
0.92	144⅛	149⅝	5½	144⅜	149⅜	5	144⅝	149¼	4⅝
0.95	151¾	156⅛	4⅜	152	156	4	152¼	155¾	3½

STROKE TABLE.

Piston Position. (Stroke = unity.)	CRANK ANGLES. (FOR ORDINARY CONNECTING ROD.)								
	RATIO 1 : 7.			RATIO 1 : 7½.			RATIO 1 : 8.		
	Forward	Return.	Diff.	Forward	Return.	Diff.	Forward	Return.	Diff.
	deg.	deg.	deg.	deg.	deg.	deg.	deg.	deg.	deg.
0.125 = 1/8	39	44 1/4	5 1/4	39 1/8	44	4 7/8	39 1/4	43 7/8	4 5/8
0.2	50	56 1/2	6 1/2	50 1/4	56 3/8	6 1/8	50 3/8	56 1/8	5 3/4
0.25 = 1/4	56 5/8	63 5/8	7	56 3/4	63 3/8	6 5/8	57 1/8	63 1/4	6 1/8
0.3	62 3/4	70 1/4	7 1/2	63 1/8	70	6 7/8	63 3/8	69 3/4	6 3/8
0.333 = 1/3	66 7/8	74 1/2	7 5/8	67 1/8	74 1/4	7 1/8	67 3/8	74	6 5/8
0.375 = 3/8	71 5/8	79 1/2	7 7/8	71 7/8	79 3/8	7 1/2	72 1/8	79 1/8	7
0.4	74 1/2	82 1/2	8	74 3/4	82 1/4	7 1/2	75	82	7
0.45	80 1/4	88 3/8	8 1/8	80 1/2	88	7 1/2	80 3/4	87 3/4	7
0.5 = 1/2	85 7/8	94 1/8	8 1/4	86 1/4	93 3/4	7 1/2	86 1/2	93 5/8	7 1/4
0.55	91 5/8	99 3/4	8 1/8	92	99 1/2	7 1/2	92 1/4	99 1/4	7
0.6	97 1/2	105 1/2	8	97 3/4	105 1/4	7 1/2	98	105	7
0.625 = 5/8	100 1/2	108 3/8	7 7/8	100 5/8	108 1/8	7 1/2	100 7/8	107 7/8	7
0.65	103 1/2	111 1/4	7 3/4	103 3/4	110 7/8	7 1/8	104	110 3/4	6 3/4
0.666 = 2/3	105 1/2	113 1/8	7 5/8	105 3/4	112 7/8	7 1/8	106	112 5/8	6 5/8
0.68	107 1/4	114 3/4	7 1/2	107 1/2	114 1/2	7	107 3/4	114 1/4	6 1/2
0.7	109 3/4	117 1/4	7 1/2	110	116 7/8	6 7/8	110 1/4	116 5/8	6 3/8
0.71	111	118 3/8	7 3/8	111 1/4	118 1/8	6 7/8	111 1/2	117 7/8	6 3/8
0.73	113 5/8	120 7/8	7 1/4	113 7/8	120 5/8	6 3/4	114 1/8	120 3/8	6 1/4
0.75 = 3/4	116 3/8	123 3/8	7	116 5/8	123 1/4	6 5/8	116 3/4	122 7/8	6 1/8
0.76	117 5/8	124 5/8	7	117 7/8	124 3/8	6 1/2	118 1/8	124 1/8	6
0.77	119 1/8	126	6 7/8	119 3/8	125 3/4	6 3/8	119 5/8	125 1/2	5 7/8
0.78	120 1/2	127 1/4	6 3/4	120 3/4	127	6 1/4	121	126 7/8	5 7/8
0.79	122	128 1/2	6 1/2	122 1/4	128 3/8	6 1/8	122 1/2	128 1/8	5 5/8
0.8	123 1/2	130	6 1/2	123 5/8	129 3/4	6 1/8	123 7/8	129 5/8	5 1/2
0.81	125	131 1/4	6 1/4	125 1/8	131 1/8	6	125 3/8	130 7/8	5 1/2
0.82	126 1/2	132 3/4	6 1/4	126 3/4	132 1/2	5 3/4	127	132 1/4	5 1/4
0.83	128 1/8	134 1/4	6 1/8	128 3/8	133 7/8	5 1/2	128 1/2	133 3/4	5 1/4
0.84	129 5/8	135 5/8	6	130	135 1/2	5 1/2	130 1/4	135 1/4	5 1/8
0.85	131 3/8	137 1/4	5 7/8	131 5/8	137	5 3/8	131 3/4	136 3/4	5
0.86	133 1/8	138 3/4	5 5/8	133 1/4	138 1/2	5 1/4	133 1/2	138 3/8	4 7/8
0.87	134 3/4	140 1/4	5 1/2	135	140 1/8	5 1/8	135 1/4	140	4 3/4
0.875 = 7/8	135 3/4	141 1/8	5 3/8	136	140 7/8	4 7/8	136 1/8	140 3/4	4 5/8
0.88	136 5/8	142	5 3/8	136 7/8	141 3/4	4 7/8	137	141 5/8	4 5/8
0.89	138 1/2	143 5/8	5 1/8	138 3/4	143 1/2	4 3/4	138 7/8	143 3/8	4 1/2
0.9	140 1/2	145 3/8	4 7/8	140 3/4	145 1/4	4 1/2	140 3/4	145	4 1/4
0.91	142 5/8	147 1/4	4 5/8	142 3/4	147 1/8	4 3/8	143	147	4
0.92	144 7/8	149 1/8	4 1/4	145	149	4	145 1/8	149	3 7/8
0.95	152 3/8	155 5/8	3 1/4	152 1/2	155 5/8	3 1/8	152 5/8	155 1/2	2 7/8

Referring to this Ratio column in the Stroke Table we obtain :—

Crank angle of the forward stroke for the 0.65 position $=112\frac{3}{8}°$.

Crank angle of the return stroke for the 0.65 position $=122\frac{1}{4}°$.

Difference between the return and forward $=9\frac{7}{8}°$.

SECOND EXAMPLE.

Stroke of piston$=45$ inches.
Ratio of crank to rod$=1:6\frac{1}{4}$.
Forward stroke crank angle$=131\frac{1}{4}°$.
Return stroke crank angle$=134\frac{3}{8}°$.
What locations will the piston occupy for these angles?

From the Stroke Table we learn that :—$131\frac{1}{4}°$ forward$=$ piston location of 0.85 the stroke and $134\frac{3}{8}°$ return$=$piston location of 0.83 the stroke—consequently :

$45'' \times 0.85 = 38\frac{1}{4}$ inches from commencement of forward stroke.
$45'' \times 0.83 = 37\frac{3}{8}$ " " " return "

ECCENTRIC AND VALVE CONNECTION.

The principle of this connection has already been illustrated by Fig. 4, its standard motion in Fig. 5, but as the latter rarely occurs in practice it becomes necessary to study the former with reference to its influence on the events of the valve motion. It has been observed that the combination is nothing more nor less, than that of a small crank with a long connecting rod, the valve will therefore move in precisely the same manner as the piston, and will have in its progress from one extremity of the travel to the opposite, like irregularities, differing only in degree. In other words, when the eccentric arrives at the positions for cut-off and lead, the valve will be drawn *beyond* its true position—measured towards the eccentric—by a distance dependent on the ratio between the throw of the eccentric and the length of its rod. Since this difficulty is corrected by *lengthening* the rod, it follows that the *width* of the port opening in one stroke, will slightly *exceed* that in the other. This is practically the only effect produced by the use of the true eccentric connection ; although strictly speaking there is besides a slight difference in the equality of the exhaust closure, yet in no case does this become sufficient to affect the general action.

Neither is the difference in the opening appreciable in stationary engines, for their ratio of eccentric throw to length of rod is usually that of 1 : 20 or 30, which gives a variation too small to influence the general admission of the steam.

It does not come within the province of this work to introduce and explain The Indicator*—that most valued friend of the Engineer, whose card ever furnishes clear and indubitable proof of the character, time and correlation of the various events taking place within the cylinder, but the Author cheerfully testifies to its many excellencies and commends it to the Reader.

---

* For a complete analysis of this instrument, its practical operation, etc., the reader is referred to Mr. Charles T. Porter's Treatise on the RICHARD'S STEAM INDICATOR, enlarged by F. W. Bacon, M. E., and published by D. Van Nostrand, New York.

PART IV.

LINK MOTIONS.

LINK MOTION.

The various mechanical devices embraced under this general term, have many strong points of resemblance and subserve a common object. By means of them, the Engineer is able at will to change the direction of the crank rotation, with only the loss of the time required for overcoming the momentum of the moving parts, and developing the like in a reverse direction. More than this simple result was not contemplated in the original discovery of the link. Subsequently, however, it was found to be capable of regulating the cut-off of the steam, so that the power could always be adjusted to the work required. This feature greatly enhanced its value, and placed the engine under the complete control of the operator.

The extreme simplicity of the parts of the link motion, has enabled it to contend successfully with all rivals, and at the present day it remains in substantially its primitive form. It is applied principally to locomotive and marine engines, where the power demanded is quite variable, and the motion at one time direct, at another reverse.

The designs may be divided into four classes:

I. The shifting link motion.
II. The stationary link motion.
III. The Allan link motion.
IV. The Walschäert link motion.

The first form was invented by Mr. Howe, in 1843, and applied to the locomotives of Messrs. Robert Stephenson & Co. It is in fact the representative link motion, which, excepting slight modifications in the mode of suspension, remains unchanged by the accumulated experience of a quarter of a century.

Simultaneous with the appearance of this motion was that of the second, the discovery of Mr. Daniel Gooch. It accomplishes perfectly analogous results, and has met with much favor throughout Great Britain and the Continent.

The "Allan" combines the characteristic features of the Howe and Gooch link motions in such a manner that the parts are more perfectly balanced, consequently it dispenses with the counter weight or spring peculiar to the former of these motions.

The Walschaert motion is extensively applied in Belgium, but probably will not receive much attention from locomotive Engineers, beyond the limits of that Kingdom, unless future designers succeed in reducing the number of its connections.

It is proposed to confine our investigation to the shifting link motion, to develop the general laws governing its action amid varied conditions, to present graphic methods for determining the proportions of the parts, and briefly to point out the general application of the same to the link motions of the other three Classes.

SHIFTING LINK MOTIONS.

A link, operated by two fixed eccentrics, forms when properly suspended an exact mechanical equivalent of the movable eccentric. Unlike the latter, however, its motion is

capable of an accurate adjustment, which practically nulli-
fies the effect of irregularities in cut-off and exhaust closure,
attributable to the angularity of the main connecting rod.

The general form in which its parts are arranged in
American locomotive practice, is clearly shown in Fig. 22.
Upon the main shaft are keyed the forward and backing
eccentrics, with their centres at F and B, so located as
to secure the most appropriate angular advance. Their
straps are bolted to the eccentric rods, and these in turn
are pinned to the "link." The slide valve is attached by
its stem to one of the rocker arms, and a "block" sur-
rounds the pin of the opposite arm, which fits the main
link and slides freely therein. The centre of the link is
spanned by a plate called the "saddle," on which is formed
the pin or stud that supports the link and eccentric rods.
This pin is embraced by a bar called the "hanger," or
sometimes the suspending or the sustaining link, from its
position and the service rendered to the motion. The
former term is preferable on account of its conciseness, and
can lead to no confusion. The opposite extremity of the
hanger is attached to one arm of the tumbling shaft. Both
arms of this shaft are rigidly secured, and form upon it a
"bell crank." The shaft itself freely oscillates on properly supported bearings, but is limited in its motion by the
action of the reversing rod. The link has been dropped
into the full gear forward, thus throwing the entire influence of the eccentric F upon the valve motion to the almost
complete exclusion of that of its mate B. By drawing back
the reversing rod and raising the link until the pin of the
other eccentric rod is brought in line with the pin of the
rocker arm, the link will be made to occupy a location appropriate to a negative crank movement (4th quarter, Fig. 9)
and *intermediate suspensions* will in like manner be pro-

ductive of *earlier* cut-off and exhaust closures. In order to clearly demonstrate that such similarity exists between these motions, it will be necessary to reduce Fig. 22 to a skeleton form like Fig. 23, and follow the journeyings of the "link arc" throughout a complete revolution of the crank.

Let the path of the main crank pin be represented by the circle E D in Fig. 23. This being divided into 12 equal parts, gives a sufficient number of positions for the purpose of tracing the motions of the link arc. The zero will be known as position No. 1, the 180° as position No. 7, and so on. Within this circle describe the path of the eccentric centres by means of the circle F B b'. This should first be divided into 12 equal parts, with F as the origin of one eccentric's motion, and again into 12 other equal parts with B as an origin, so that when the crank moves from position No. 1 to 3 the new positions f^3 and b^3 of the two eccentrics may be instantly found, and the same with other locations. The original positions F and B are of course laid off with the angular advance due to the proposed maximum cut-off. At the distance C t from the centre of the shaft erect the perpendicular T t and locate T the fixed centre of the tumbling shaft. T h will represent the arm which supports the link through its hanger and h h' h'' the arc described by this arm. A second perpendicular at the distance C A will contain the point R, the centre of the rocker shaft, whose arm R A sweeps the arc r A r. The motion of the upper arm, being merely the reverse of the lower, need not be considered, and so long as the angular advance is properly located no error can arise from the omission. In the motion of the lower arm there are five locations of vital importance, viz: one at which the exhaust of the valve opens or closes, two appropriate to the lead at full gear of the link, and two at which cut-off takes place or the valve closes its ports. The 1st is evidently the normal position R A of the rocker, the 2d R d, R d', that in which the rocker pin is drawn aside a distance A d equal to the sum of the lap and lead, and the 3d R l, R l' corresponds with a removal A l equal to the lap. Hence, so far as the slide valve is concerned we can confine our attention to the motion of the rocker arm pin upon the arc r r. The five positions in question can be distinctly located by sweeping a circle d d', with a radius equal to the lap plus the lead of the valve, around the exhaust point A, and inscribing a second circle l l' with a radius equal to the lap of the valve. Then the four points in which these circles intersect the arc r r will give the 4 positions of the pin corresponding with the lead and cut-off positions of the valve, and the centre of these circles will give the exhaust closure positions. As these locations will be constantly referred to in the sequel, it should be remembered that the "lead circle" d d' fixes those points on the arc r r which the pin of the rocker arm must occupy when the valve has a given lead; and that the "lap circle" l l' locates the positions of the same pin for the moments at which the steam ports are closed against the admission of steam to the cylinder.

Our next duty will be to reduce the link to its simplest form.

It appears on examination that the rocker pin is entirely subject, in its motion, to the guidance of the link arc, and that this arc swept with a radius C A is rigidly connected with three moving points, viz. the saddle pin, and the two eccentric rod pins. In following the motion of the link arc, the connection of the parts can best be maintained by the use of a template, cut from white holly veneer or other hard wood and shaped like L L in Fig. 23, upon which are

made ∨ shaped incisions for locating the points f, S and b
of the pins.

We are now prepared to find position No. 1 of the link
corresponding with No. 1 of the crank. Of course when
the crank is at the zero the steam port should be opened an
amount equal to the lead of the valve. The rocker arm
therefore will occupy the position R d, and the point d lie
in the link arc. Since the eccentric centres F, B are found
in a line perpendicular to the central line of motion, and
the eccentric rods are of equal length, the link must occupy
a nearly perpendicular position. Place the template so that
its arc coincides with the point d and mark the point f upon
the paper, then the distance from F to f will equal the
length of the eccentric rod. With this length as a radius
describe about F as a centre the small arc f g, likewise
with B as a centre describe the small arc b h. Apply the
template to these arcs so that the points f and b shall be
found in them and the point d on the link arc n c d, after
which, draw the link arc on the paper and we obtain position No. 1 of the link. With the saddle pin S as a centre
and the length of the hanger h S as a radius, the position T
h of the tumbling shaft arm is readily found for full gear
of the link and conversely the arc c S c is fixed along which
the saddle pin must travel during the revolution of the
crank.

The preparatory stages of our solution are now complete,
the link motion of Fig. 22 has been reduced to its skeleton
form and the first position of the link located. Our next
step is to follow the link arc during its journeyings in a
single revolution of the crank. Suppose then, the crank
is made to occupy position No. 2, the eccentrics will be
carried forward from F and B to f^2 b^2. Since the length
of their rods remains unchanged the arcs f g, b h, will be

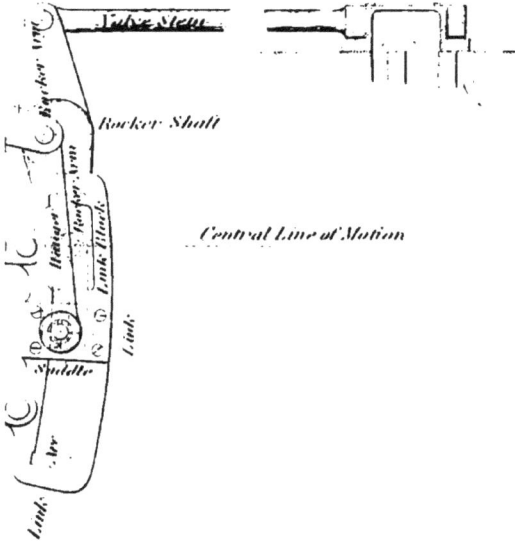

Valve Stem

Rocker Shaft

Rocker Arm

Hanger

Rocker Arm

Link Block

Saddle

Link

Link

Central Line of Motion

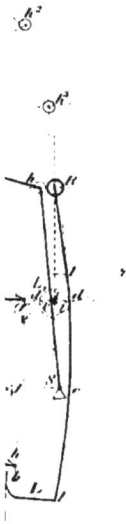

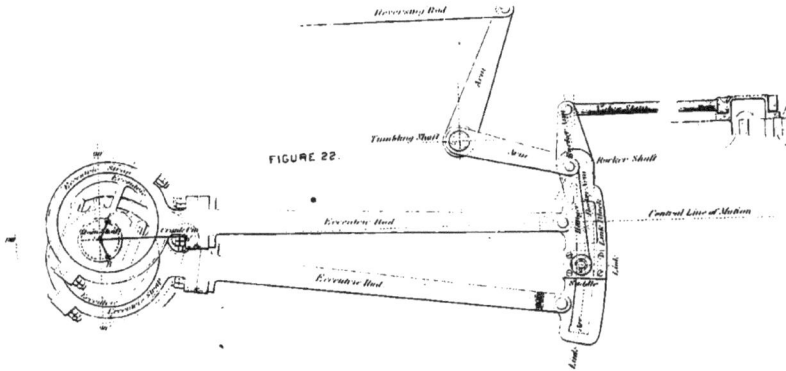

FIGURE 22.

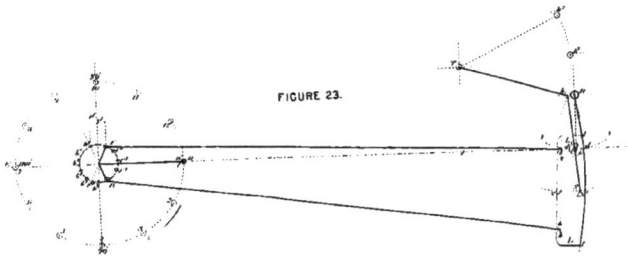

FIGURE 23.

Fig. 24.

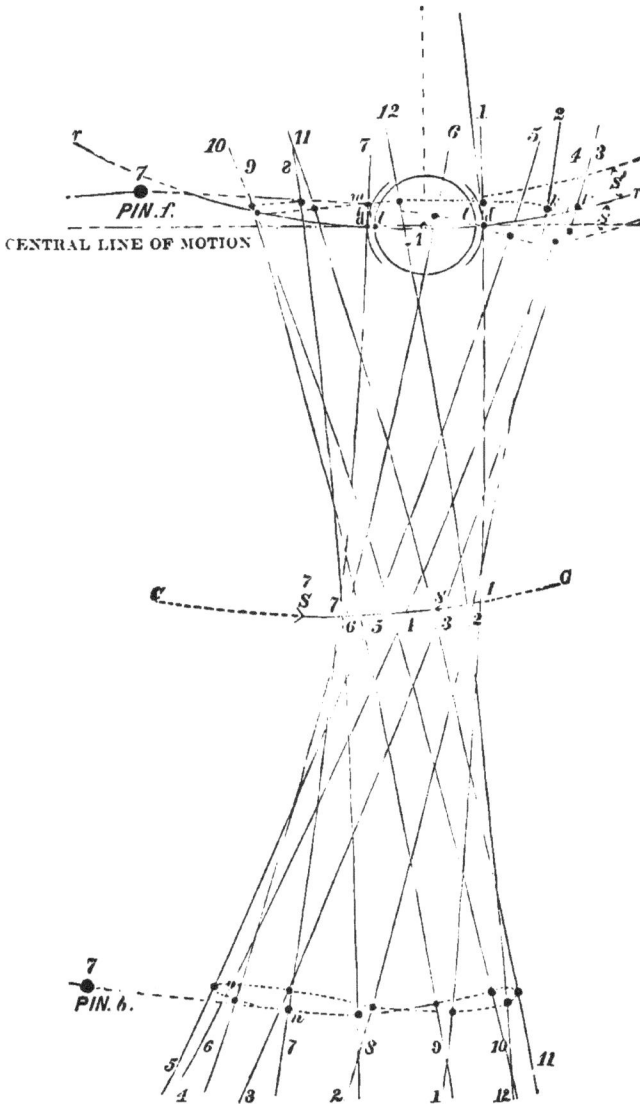

removed from their first position and the link template will
follow them with its points *b* and *f*. The only restraint

7

upon the course of this template is that the point S must travel on the hanger arc $c\,c$. If therefore we describe new arcs about the centres $f^2\,b^2$, and adjust the template so that f and b shall be found in those arcs and S in the arc $c\,c$ there will result a new link position with its arc standing like 2 2 in Fig. 24 and intersecting the rocker pin arc $r\,r$ at a point k. But as the rocker pin necessarily follows the course of the link arc it will by this change be drawn aside from d to k, consequently the steam port will be opened wider by the extent of the horizontal measurement of this distance. In like manner when the crank is carried to position No. 3 the link arc will be removed to 3 3 as in Fig. 24, and the rocker pin to V, producing thereby a still wider opening of the steam port. The same process applied to the remainder of the 12 crank positions will give the other locations of the link arc (as in Fig. 24) for the full gear of the link. Now observe that the link position 3 3 produces the widest opening of the steam port, and as the crank advances to 4 and 5 this opening grows less and less, until between 5 and 6 the rocker pin reaches the point l, where the steam is finally cut off. During its further progress expansion goes on and at last when A is attained the exhaust opens and the steam escapes. At position No. 7 (the 180° location of the crank) the link arc is brought again in contact with the lead circle and a like process is repeated throughout the return stroke.

A duplicate set of link arc locations, might readily be obtained by raising the link to the full gear back position and a similar set for the mid gear, but an examination of the one just found will develop the character of the motion.

Besides the qualities possessed in common by the two motions, the link has that of adjustability, a very important feature, and one which specially characterizes it. As the tendency of the connecting rod angularity in a direct acting engine is to produce a *later* cut-off on the forward stroke than the amount required, and since with the link the cut-off in either stroke depends on its degree of elevation or depression ; it follows that if we suspend the link in such a manner as to cause a suitable elevation for the forward stroke, the result will be a perfectly equalized motion for the gear in question. And again if the equalization be made applicable to all gears, then the link may be suspended at *any* point between the full forward and full back *without* an appreciable inequality appearing between the cut-offs or the exhaust closures of either stroke.

But a practical difficulty here arises ; the link block moves upon a fixed arc *r r* while the link rises and falls, consequently for each revolution of the crank *the link will slip* back and forth a certain distance on its block. Should this slip be excessive in any particular gear and the engine run a long time in this gear, the faces of the link would become worn, "lost motion" would ensue and the delicate action of the parts would be destroyed.

Hence in planning a serviceable link motion it is necessary to reduce the slip of the link to its smallest value, consistent with the equalization of the motion, and in *marine engines* to even sacrifice the equality of the cut-offs to the *reduction of the slip*. In Fig. 24 the motion of the two fixed points (*m* and *n*) on the link have been traced in looped curves. The upper of these, shows to what extent the point *m* falls below and rises above the arc *r r*, giving a slip equal to the distance S plus S'.

It is important to observe that the magnitude of the slip grows smaller and smaller as the link block draws nearer to the point of suspension, because this fact indicates that the stud of the saddle should be placed—when a minimum value of the slip is required at a certain *point* of suspension—as *nearly over* such point as possible.

CONNECTION OF ECCENTRIC RODS.

The variable character of the lead opening in a shifting link motion depends upon the manner in which its eccentric rods are attached, and its magnitude depends on the length of those rods. The force of this remark will appear from an examination of Figures 26 and 27. In both instances, *the eccentric centres lie between the centre of the shaft and the link*, while the latter for sake of simplicity has been made to act directly on the valve. The No. 1 position represents the mid gear, and No. 2 the full gear forward of the link. If under these conditions the eccentric rods be *crossed* as in Fig. 26 the lead opening will *decrease* from the full to the mid gear of the link, where the motion may even be without lead.

But with the *open* rods of Fig. 27 the lead opening

FIG. 26.

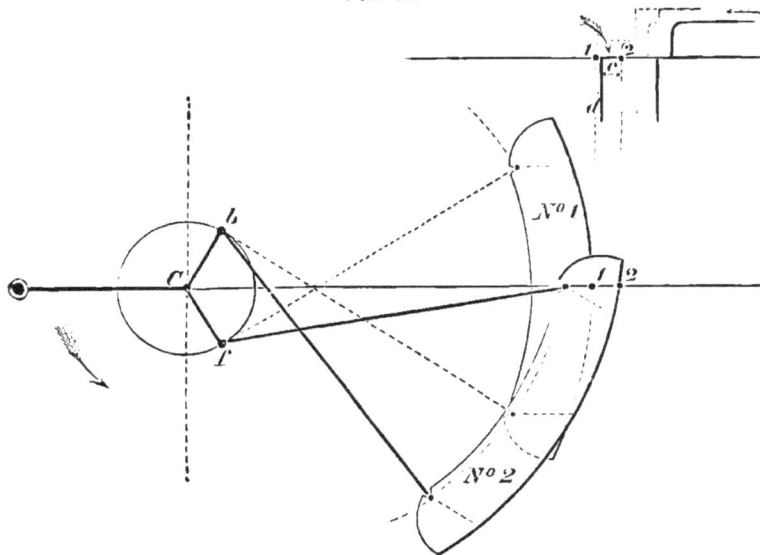

FIG. 27.

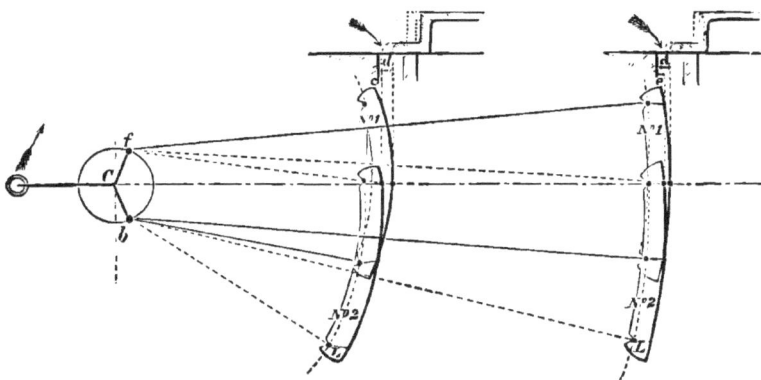

increases from the full to the mid gear, and the rapidity of
this increase, for a given link, depends directly upon the
length of the rods ; hence with a given mid gear lead open-

ing that for the full gear will be determined mainly by this length. Excepting the case of valves having an independent cut-off (Part V.) the rods are seldom crossed as in Fig. 26, yet there are good reasons for believing that many instances exist in which the arrangement might be adopted with good results. It is also possible, with such a motion, to stop the engine by placing the link in the mid gear; but this can never be done with a motion like Fig. 27, whose valve is invariably opened a certain amount in the mid gear. The extremes of mid gear lead opening in locomotive practice are $\frac{1}{4}$ and $\frac{1}{2}$ an inch, but the more common value is $\frac{3}{8}$ inch; while the full gear lead varies between $\frac{1}{16}$ and $\frac{3}{16}$ inch, governed principally by the length of the eccentric rods.

With the stationary link the lead opening remains unaltered by changes in gear; so that if $\frac{3}{8}$ inch be assumed as the proper amount for the full gears, the motion will retain this lead for all gears between these extremes and the mid gear. This peculiarity is not inherent with the stationary link, since many *shifting link motions* may be arranged with a *Constant Lead* for the various gears of *one* direction of the motion. Take, for example, the motion shown in Fig. 22, in which the angular advance of each eccentric equals 21° and the lead enlarges from $\frac{1}{8}''$ in the full, to $\frac{3}{16}''$ in the mid gear. By imparting an angular advance of 31° to the eccentric F, while that of B remains unaltered, the lead opening becomes *constant* for all points between the full gear forward and the mid gear, and diminishes from $\frac{3}{16}$ inch in the mid gear to $\frac{1}{8}''$ in the full gear back. *Vice versa* for a change in the angular advance of the eccentric B.

PRACTICAL OBSERVATIONS.
(BASED ON FIG. 22.)

I. The tumbling shaft must be located at such a distance above or below the central line of motion, that neither eccentric rod can *strike against* it when the link is moved from one full gear to the other. Special cases may arise that demand a curvature of the eccentric rod, but the practice in general should be discountenanced.

II. The hanger must be of such a length that the extremity of the link will not conflict with the tumbling shaft arm in either forward or back gear. The length of the tumbling shaft arm is usually equal to or greater than that of the hanger.

III. If the link cannot be placed in full gear back, owing to the arrest of its tumbling shaft arm by the boiler or other opposing object, either the tumbling shaft must be removed and located *below* the link motion, or the rocker must be lengthened in order to depress the central line of motion and with it, the entire motion. When the latter expedient is resorted to, a change should be made in the relative positions of the rocker arms, for the purpose of preserving the identity of their motions. The proper inclination W of the arms is found by describing a circle $r\,t\,t\,r$ (Fig. 28) tangent to the central line of the valve stem, or a line sufficiently above the same to equalize the vibration of the stem, and the central line of the motion. Radial lines from the points of tangency will then give the relative positions of the arms.

This method of correction is preferable to the former in respect to the symmetry of the motion, because the greater the length of the rocker arm, the less will be the vibration of the valve stem, as well as the slip of the link block.

Fig. 29.

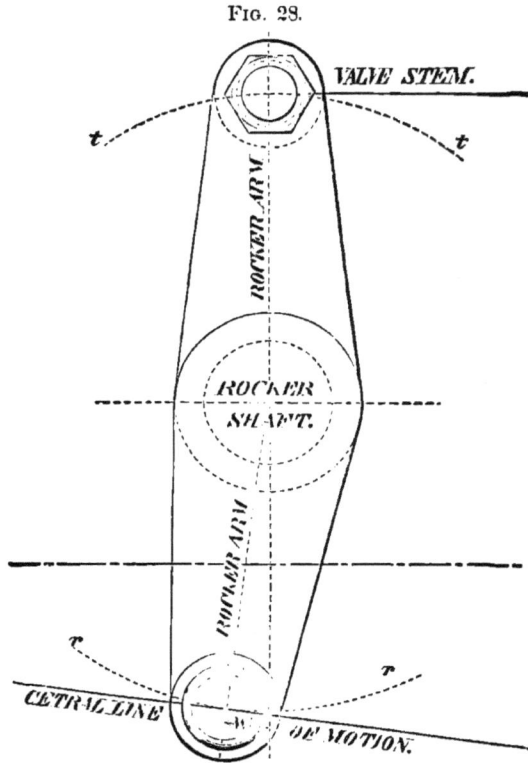

IV. So long as the angular advance of the eccentrics is laid off from a line at right angles to the central line of the link motion, the latter can be arranged at any inclination to the piston motion, without affecting the action of the link. These central lines were made to coincide in Fig. 22, merely for the purpose of simplifying the investigation, whereas they might have formed with each other any angle whatever (see Fig. 52, Part V).

GENERAL DIMENSIONS.

In ordinary locomotive practice the dimensions of the various parts range between the following extremes:

Ratio of crank arm to connecting rod $1:5\frac{1}{2}$ and $1:8$.

Travel of valve 4 to 6 inches.

Maximum cut-off from $\frac{3}{4}$ to 0.92 of the stroke (generally $= \frac{7}{8}$).

Mid-gear lead $\frac{1}{4}''$ to $\frac{3}{8}''$, usually the latter.

Full gear (dependent on length of rods) $\frac{1}{10}''$ to $\frac{3}{16}''$.

Radius of link 3′ 6″ to 6 ft.

Distance between eccentric rod pins 10″ to 14″.

Pins back of link arc $2\frac{1}{4}$ to 3 inches.

Saddle-stud back of arc 0″ to $1\frac{1}{2}$ inches.

Stud above central line of link 0″ to $2\frac{1}{2}''$.

Length of hanger 12 to 20 inches.

Length of tumbling-shaft arm 14 to 22 inches.

Length of rocker arm 8 to 11 inches.

The following special dimensions, collated by the Master Mechanics' Associations, are indicative of the prevailing practice on thirty-five of the railroads of our country:

Number of Roads and Class of Locomotives.	Travel of Valve.	Outside Lap.	Lead in full Gear.	Inside Lap.
	in.	in.	in.	in.
25 Roads use on their Express Pass. locomotives	5	$\frac{7}{8}$	1–10	$\frac{1}{8}$
6 " " " " "	$4\frac{3}{4}$	$\frac{3}{4}$	$\frac{1}{8}$	1–16
4 " " " " "	5	$1\frac{1}{4}$	$\frac{1}{8}$	$\frac{1}{4}$
20 " " Accomd. " "	5	$\frac{3}{4}$	1–10	$\frac{3}{8}$
10 " " " " "	$5\frac{1}{2}$	$\frac{7}{8}$	1–16	1–16
5 " " " " "	$4\frac{1}{2}$	$\frac{5}{8}$	$\frac{1}{8}$	3–16
19 " " Freight locomotives	5	$\frac{3}{4}$	1–10	1–16
11 " " " "	$4\frac{1}{2}$	$\frac{5}{8}$	1–16	$\frac{1}{8}$
5 " " " "	$4\frac{3}{4}$	$\frac{1}{2}$	1–10	3–16

GENERAL PRINCIPLES.

OF THE GEOMETRIC SOLUTION.

A cursory examination of the link motion might naturally lead to the conclusion—from the simplicity of the parts and the strong resemblance existing between their action and the single eccentric's—that the theory of the latter being perfectly comprehended, but little difficulty would attend the work of assigning proper proportions to the former.   Such an inference, however, would not be strengthened by a closer inspection, much less sustained by an intelligent effort to accomplish a solution.   The reason for this fact lies not only in the multiplicity of the parts, but also in the conflicting character of the elements that constitute a perfectly *equalized* link motion.

The requirements of such a motion are—*perfect equality* of cut-off, of exhaust closure, lead opening and maximum port opening, together with absence of block slip, between the forward and return stroke of the piston for *every* suspension of the link from full gear forward to full gear back. Such theoretical excellence is *absolutely* impossible with the ordinary type of link motion, and efforts made to attain the same must necessarily result in failure.

But good practical qualities *may be* obtained by *sacrificing* the *non-essential* to the *essential* points of the motion.   The action of the connecting rod on a link motion, may justly be compared to the distorting effect of pressure exerted upon one point of a symmetrical india-rubber ball, producing thereby a temporary concavity.   This it is true can be removed by an *even* application of additional pressure to the adjoining parts, but the ultimate effect will be a bulging out of the central portion, and the symmetry can

alone be restored by withdrawing *all* pressure. Just so
with the link motion, the angularity of the rod tends to render one or more events of the motion unequal in the opposite strokes of the piston, and should it appear more desirable to preserve certain ones of these than others, we must
purchase their equality at the *expense* of the latter. Reducing the angularity of course diminishes its disturbing
effect, hence in departments like locomotive engineering,
where much attention is bestowed on the equalization of the
motion, crank and connecting rod ratios of 1 : 7 or 8 obtain ;
while in marine engineering ratios of 1 : 4 or 5 are common.

The subject of preserving the equalities of cut-off and
exhaust closure at the expense of lead and port openings
has been considered already. It will only be necessary to
examine it here with reference to the mid gear. At this
point the port and lead openings attain their minimum
value, which being much less than the 0.6 or 0.9 port opening required for perfect admission, tends to reduce the pressure of the steam by wire-drawing, and if these openings
vary, unequal powers will be applied in opposite strokes.
Consequently the *mid gear, lead and port openings* must
have *equal values* in both strokes, however irregular they
may be in the full gears. No fixed limit can be assigned
to the slip of the link on its block, but the amount allowa-
ble under different conditions will readily be determined by
the judgment of the Engineer. In every case, the main object is to reduce the slip to a minimum value for that gear
in which the engine will be most frequently operated.

LINK No. I.

In designing an engine, as a
general thing, no particular part
can be isolated, its proportions
assigned, and its details worked
out regardless of the conditions
inevitably imposed upon it by
the character of the adjoining
parts; but rather, trial dimensions
must be affixed, their adaptability
tested and modified by circumstances, and finally all must be
unfolded and developed in perfect harmony. When the subject of scheming the link motion
comes in order, we find that peculiarities of detail have already
fixed the ratio of the crank arm to the connecting rod,
have pointed out a convenient location for the rocker shaft
and have more or less circumscribed the boundaries of the
entire motion.

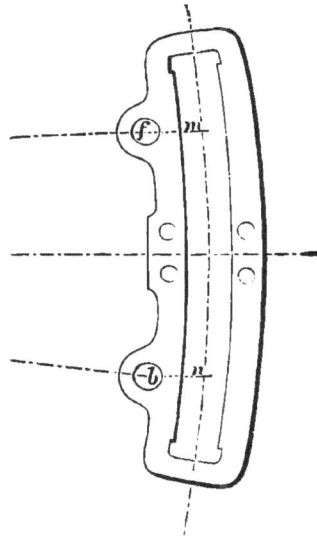

Since methods of construction are always most intelli-

gibly presented when the mind is able to follow their operation in the solution of a practical example, we will take
for illustration of our method the following dimensions :

Ratio of crank to connecting rod $= 1 : 7\frac{1}{2}$.
Eccentric circle diameter $= 5\frac{1}{2}$ inches.
Maximum cut-off $= 0.92$ stroke.
Rocker from shaft $= 49\frac{1}{4}$ inches.
c. to c. of eccentric pins $= 13$ inches.
Pins back of link arc $= 3$ inches.
Mid-gear lead $= \frac{3}{8}$ inch.

To find lap, full-gear lead, point of suspension of link
and location of tumbling shaft.

———————

Spread upon a long drawing board, or table, two sheets
of paper large enough to contain figures similar to 29 and
31 when drawn on the Full Scale. The one will be used for
locating the various important positions of the eccentric
centres ; the other, for the journeyings of the link and its
point of suspension, their centres should therefore be separated by the proposed distance between the shaft and
rocker.

Stretch a fine thread tightly across both papers in order
to locate the right line E C D A, which constitutes the
Central line of Motion.*

Describe about the point C as a centre the eccentric
circle E F D, with a radius equal to the throw of the eccentric. Then, with the points of intersection E and D as
centres describe with an assumed radius equal arcs intersecting at G and H and erect the perpendicular G C H to
represent the neutral positions of the eccentrics. From this

———————

* The use of the T square should be avoided in all of the constructions.

line lay off an angular advance appropriate to the desired
cut-off. This may be found in the subjoined Table:

Cut-Off.	Angular Advance.	Return Stroke Maximum Cut-Off Angle.
0.75 = ¾	28 degrees.	124 degrees.
0.8	25 "	130 "
0.84	22 "	136 "
0.875 = ⅞	20 "	140 "
0.9	17 "	146 "
0.92	16 "	148 "

In the present case the advance equals 16°, which laid
off, by means of a protractor, from the line C G, determines
the position F of the forward motion eccentric when the
crank stands at the zero. In like manner B might be found,
but it will always prove more convenient and accurate to
take in a pair of dividers the distance between F and the
point of intersection of the circle with the line G C and then
prick off from the line G H the points b, f, B. We thus
obtain the two positions F and B of the forward and backing eccentrics when the crank stands at the zero, as well as
their new ones f and b for the 180° location of the crank.

It is well known that the inequality of the crank angles
attains its maximum value at the ½ stroke of the piston,
hence the importance of examining the link motion with
special reference to the ½ stroke cut-off. Although appropriate
angles for the crank have been furnished in the Stroke Table,
it is thought best for facility of reference to here reproduce
them in a more compact form.

Ratio of Crank to Connecting Rod.	Crank Angles for Half-Stroke Position of the Piston.	
	Forward Stroke.	Return Stroke.
1 : 4	$82\frac{7}{8}$ degrees.	$97\frac{1}{8}$ degrees.
1 : $4\frac{1}{2}$	$83\frac{3}{4}$ "	$96\frac{1}{4}$ "
1 : 5	$84\frac{3}{8}$ "	$95\frac{5}{8}$ "
1 : $5\frac{1}{2}$	$84\frac{7}{8}$ "	$95\frac{1}{8}$ "
1 : 6	$85\frac{3}{8}$ "	$94\frac{5}{8}$ "
1 : $6\frac{1}{2}$	$85\frac{5}{8}$ "	$94\frac{3}{8}$ "
1 : 7	$85\frac{7}{8}$ "	$94\frac{1}{8}$ "
1 : $7\frac{1}{2}$	$86\frac{1}{4}$ "	$93\frac{3}{4}$ "
1 : 8	$86\frac{3}{8}$ "	$93\frac{5}{8}$ "

If the ratio of crank to connecting rod be that of 1 : $7\frac{1}{2}$ the two eccentrics will advance $86\frac{1}{4}°$ from F and B while the piston travels to its forward $\frac{1}{2}$ stroke location, and $93\frac{3}{4}°$ from *f* and *b* for the return stroke.

A very convenient way of locating these four points is to lay off from C F by means of a protractor, the point $\frac{1}{2}$ *f* distant $86\frac{1}{4}°$, and *f*$\frac{1}{2}$ distant $93\frac{3}{4}°$ from *f*C. Then with a pair of dividers prick off these points at equal distances on the opposite side of E and D giving thereby the two other points $\frac{1}{2}$*b* and *b*$\frac{1}{2}$. In the nomenclature here adopted the letter *f* refers to that eccentric which produces a forward or positive motion of the crank, while *b* always designates the eccentric for the back or reverse motion.

When a fraction is prefixed to either of these letters, it signifies some forward stroke position of the piston with the link in the forward gear, but if it follows the letter a return stroke of the piston with the link in the same gear. Thus, the $\frac{1}{2}$*f* represents the position of the forward eccentric when the link is in the forward gear and the piston has advanced to its forward $\frac{1}{2}$ stroke location, and *f*$\frac{1}{2}$ represents the same when the piston has attained its $\frac{1}{2}$ stroke return position. Having accurately located these eight important positions

G

16°

b   F

Eccentric Rod.

86¼

b

C'   Central Line of Motion.

F   C   D

a

b

83¾   a

c   B.

Eccentric Rod.

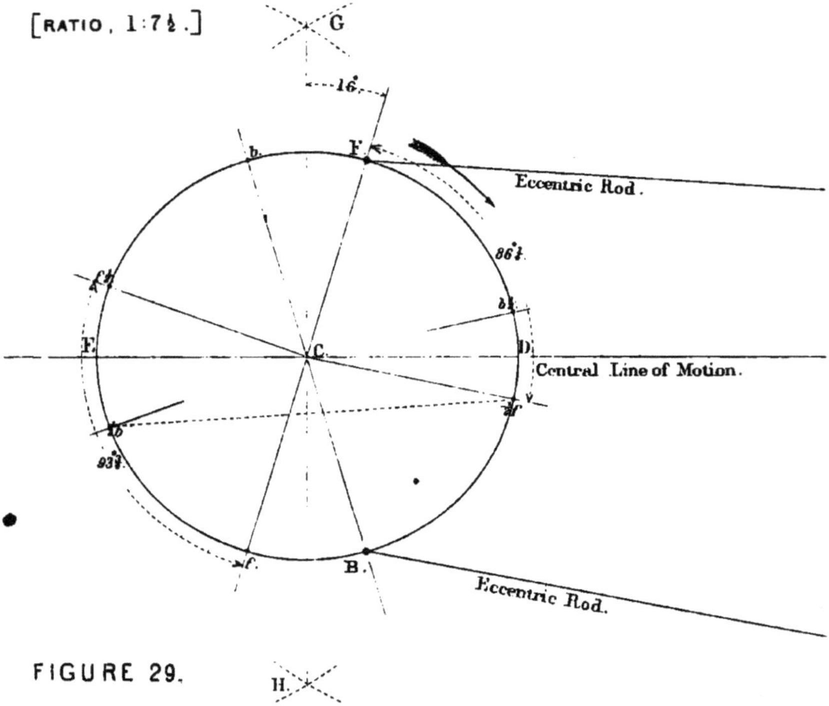

FIGURE 29.   H.

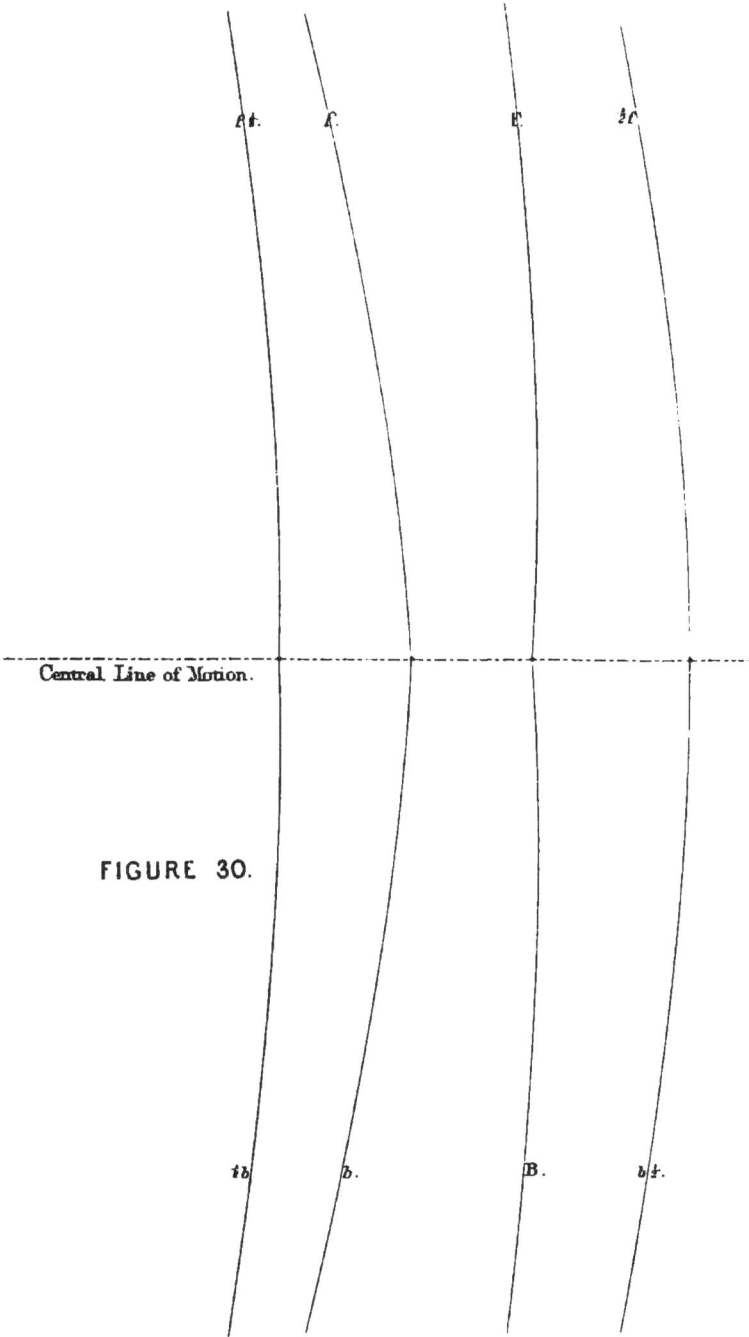

Central Line of Motion.

FIGURE 30.

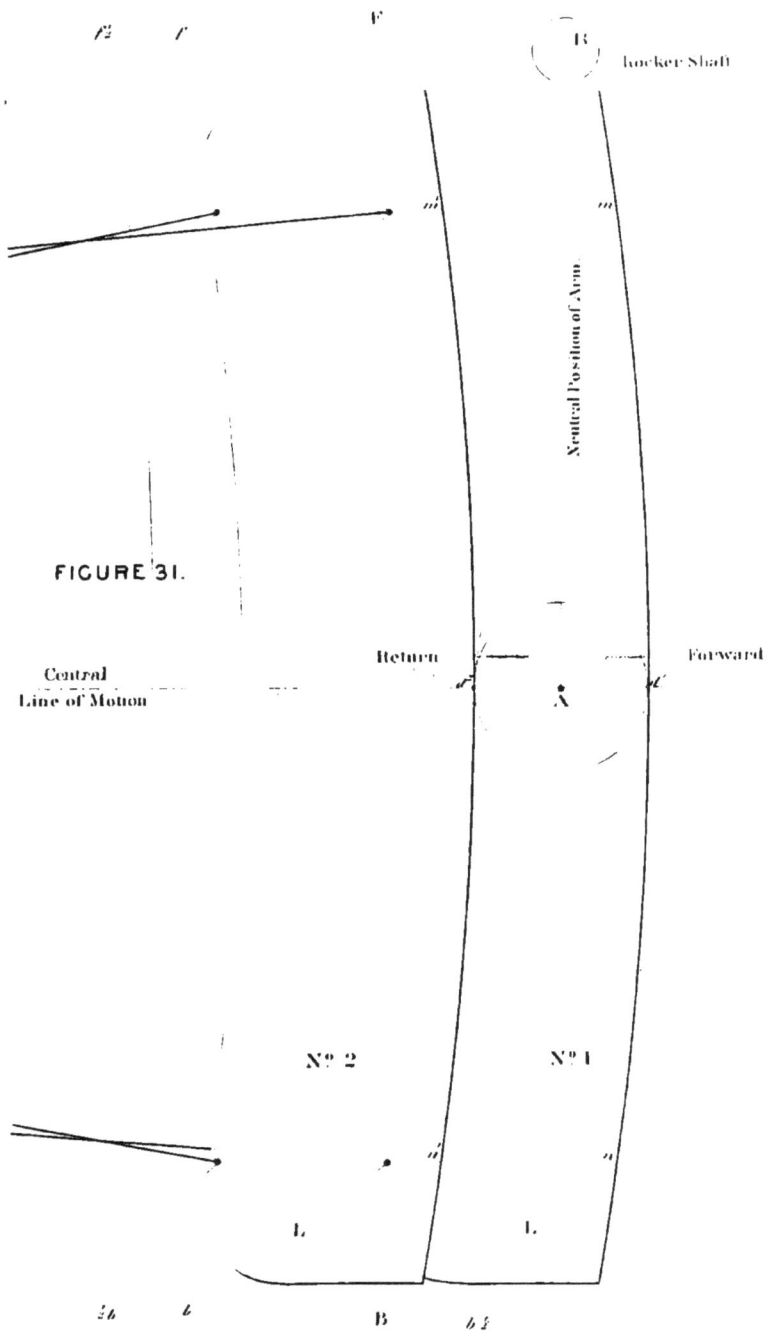

FIGURE 31.

Rocker Shaft

Neutral Position of Arm

Central
Line of Motion

Return   Forward

A

No 2   No 1

L.   L.

B

of the eccentrics, we pass to the other sheet of paper and
trace their influence on the proportions of the link attachments.

Since the assumed distance of the rocker from the shaft
is $49\frac{1}{4}''$ and the eccentric rod pins are withdrawn $3''$ back of
the link arc, the length of each rod will equal $49\frac{1}{4}''-3''=46\frac{1}{4}''$.
Adjust a pair of beam compasses to strike arcs of $46\frac{1}{4}''$
radius. Step the needle point successively in the eight locations of the eccentric's centres just found, and sweep from the
central line of motion the same number of indefinite arcs (as
shown in Fig. 30) upon which the eccentric rod pins must
inevitably travel for the four given positions of the crank
arm.

Next, from a piece of white holly veneer cut a template L
(Fig. 32) having a link arc of $49\frac{1}{4}''$ and with $\vee$ incisions, $13''$
apart and $3''$ back of the arc to represent the location of the
eccentric pins, and draw upon the same the three parallel
lines $f\,m,\,j$ S, $b\,n$; making j S lie midway between f and b
as well as perpendicular to a line joining these points. We
are now prepared to trace the journeyings of the link arc.

I. To find the mid-gear travel.

For this purpose place the template in the mid gear
positions No. 1 and 2 (Fig. 31) with its eccentric pins on the
arcs F, B, f, b, and mark the points d', d^2 in which the link
arcs intersect the central line of motion. Locate these
points permanently by describing a circle through them,
having its centre A in the central line. This point gives us
the *true* location for the rocker shaft, the distance from the
main shaft being equal to C A instead of $49\frac{1}{4}''$.[*] Through
A. therefore, erect a perpendicular A R to the central line of
motion and on it locate the centre R of the rocker shaft.

---

[*] Their difference is always so trifling that the rocker box may readily
be moved the proper amount and much difficulty of construction be thereby
avoided.

II. To find the lap of the valve.

From d^1 and d^2 lay off the mid-gear lead opening of $\frac{3}{8}$ inch towards A, and permanently locate the positions thus found by sweeping about the centre A a second circle $l\,l'$. But since the mid-gear travel invariably equals the sum of the laps plus the mid-gear lead openings, the diameter of the circle $l\,l'$ will equal the sum of the laps, consequently the simple lap of the valve must equal its radius A l or A l', Fig. 32.

The following Table will aid the designer in the selection of a suitable lead opening after the mid-gear travel has been determined. For since the value of the lead angle may range between about 30° and 40°, the widths of the openings will be those found in the Table. Of course much latitude is here allowed to the exercise of individual judgment, for the subject demands it. Observe, the larger travels are only found on marine engines:

MID-GEAR DIMENSIONS.

Mid-gear Travel.	LEAD OPENING FOR A LEAD ANGLE OF—		
	30°.	35°.	40°.
1 inch.	⅛ inch.	3/16 inch.	¼ inch.
1½ "	3/16 "	¼ "	5/16 "
2 "	¼ "	3/8 "	7/16 "
2½ "	5/16 "	7/16 "	9/16 "
3 "	3/8 "	½ "	11/16 "
3½ "	7/16 "	9/16 "	13/16 "
4 "	½ "	11/16 "	15/16 "

III. To find position of the stud for equal cut-offs at the ½ stroke of the piston.

Place the template in position No. 3, with its eccentric pins on the forward ½-stroke elements $\frac{1}{2}f\,\frac{1}{2}b$, and its link arc in contact with the lap or cut-off point l. Then mark upon the paper the position j S occupied by its central line

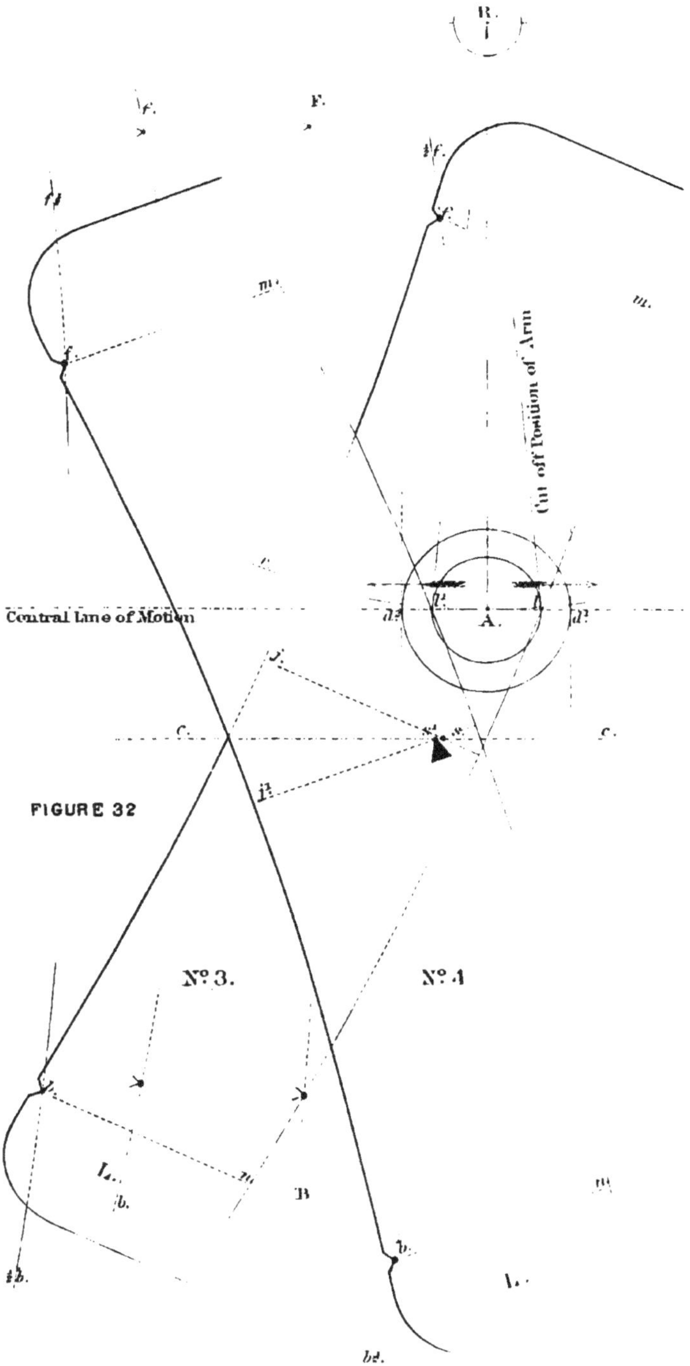

R.

F.   F.   F.

Cut off Position of Arm

m.   m.

Central Line of Motion

A.

FIGURE 32   C.   c.

No. 3.   No. 4

B

L.
b.

b.   b.

b.

together with a portion of the link arc. Next, place the
template in the position No. 4, with its pins on the arcs *f* ¼,
b ½ and link arc over the other cut-off point *l'*, after which
mark on the paper the second position, *j'* S', of the link centre line. Having decided to suspend the link centrally, the
point of suspension must be found on the line *j* S, and con-
sidering the manner in which it hangs from the tumbling
shaft it is evident that for a short distance the stud will
practically move along some straight line *c c parallel* to the
central line of motion. The point of suspension therefore
must reside in the central lines *j* S, *j'* S', must be equally
remote from the link arcs and at such a distance that a line
drawn through the two points will prove parallel to the central line of motion. The only two positions satisfying such
conditions are S and S', found by trial distances laid off
with a pair of dividers from the two link arcs. Having secured the proper distance for the stud, fix it permanently,
by making a V incision in the link template ; for as our
subsequent study of the link will be intimately associated
with the motion of this point, it is important to be able to
mark its position for other gears of the link.

The inequality of the crank angles for different positions
of the piston attains its maximum value at the ½ stroke and
gradually fades out at the extremities, and since we have
equalized the cut-off for the ½ stroke, it only remains to perform the same office for the maximum cut-off before we
practically equalize the motion for ALL *intermediate gears*
between the full and mid gears. Our next step, therefore,
will be to return to Fig. 29 and map on it the four positions
of the eccentries for the maximum cut-off. The third column of the Angular Advance Table gives the maximum
cut-off angle for the *return* stroke, which in the present in-
stance=148°, and the STROKE TABLE shows that the forward

stroke angle is $4\frac{1}{4}°$ less than the return, or=$143\frac{3}{4}°$. With these angles known, the four positions for the maximum cut-off may be readily laid off with a protractor, but the nature of the case enables us to present a more rapid solution. From C F (Fig. 33) lay off an angle of $143\frac{3}{4}°$. This

Fig. 33.

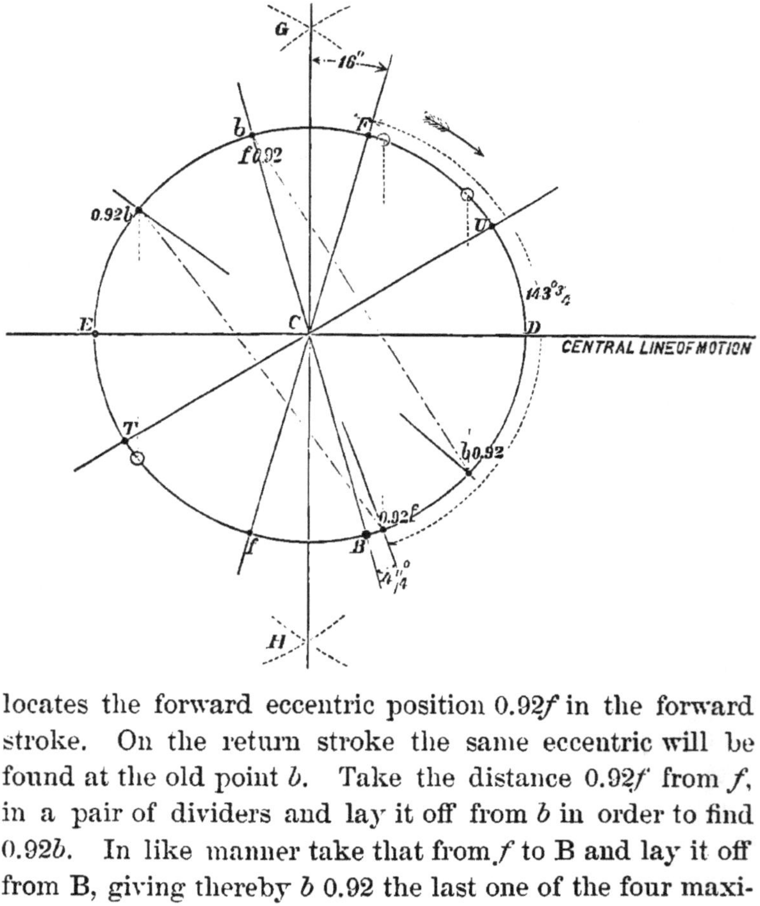

locates the forward eccentric position 0.92f in the forward stroke. On the return stroke the same eccentric will be found at the old point b. Take the distance 0.92f from f, in a pair of dividers and lay it off from b in order to find 0.92b. In like manner take that from f to B and lay it off from B, giving thereby b 0.92 the last one of the four maximum cut-off points sought. With these points as centres and the length of the eccentric rod as radius sweep indefi-

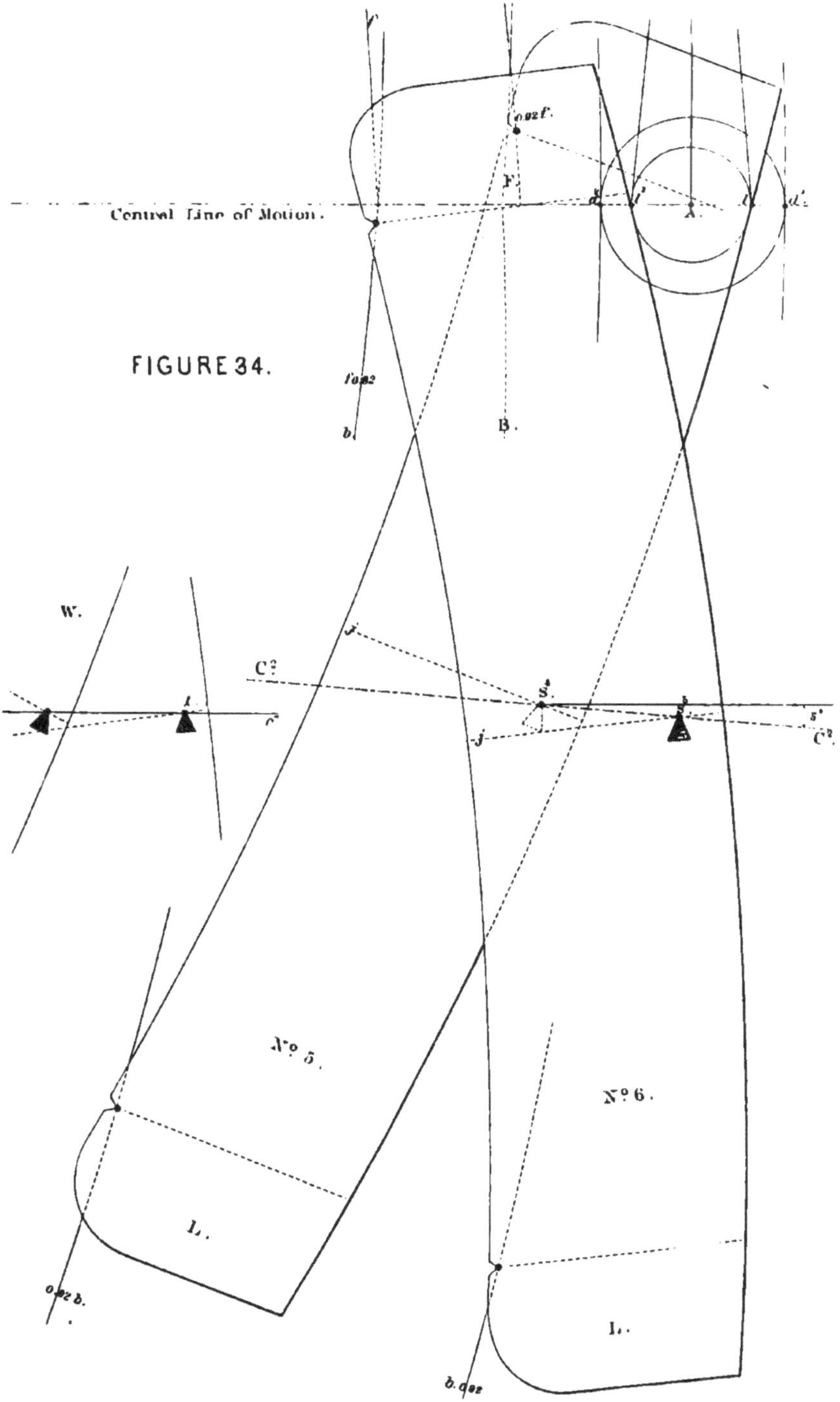

Central Line of Motion.

FIGURE 34.

W.

C²

No 5.

No 6.

L.

L.

nite arcs (as shown in Figure 34), on which the link template may travel in the full gear.

IV. To LOCATE THE TUMBLING SHAFT FOR ACCOMPLISH-ING AN EQUALIZED CUT-OFF IN ALL GEARS.

Slide the link template with its eccentric rod pins on the elements $0.92\,f$; $0.92\,b$, until its link arc comes in contact with the lap or cut-off point l (position No. 5) and mark the point S^4 occupied by the stud.

Again, slide the template on the return stroke elements $f\ 0.92$; $b\ 0.92$ until its link arc is in contact with the other lap point l', and mark the stud position S^5. Join the points $S^4\ S^5$ by a line $c^2\ c^2$, which is found to have an inclination to the central line of motion of about $5°$ instead of parallelism* as with $c\ c$.

By projecting the eccentric position points to the opposite side of their circle, sweeping indefinite elementary arcs with the eccentric rod as radius, and applying to them the link template, a corresponding set of stud locations S^2, S^3, S^6, S^7 (Fig. 35) may be found for equal cut-off in the back gear. But such efforts are uncalled for in the class of motions just described, because their back motion will be a precise counterpart of their forward motion, consequently the latter may be reproduced from the former, as in Figure 35.

Having thus determined 8 positions of the centre of suspension for equal $\frac{1}{2}$ strokes and maximum cut-offs, it only remains to sustain the hanger in such a manner that for the different elevations it will sweep arcs passing through ALL of these points. Arcs of intersection formed with an assumed length of hanger as radius and these points as cen-

---

* Parallelism might be secured by moving the stud S to within a distance t of the link arc (see Fig. W), but such a change would destroy the equality of the $\frac{1}{2}$ stroke cut-offs.

tres will locate the points h and h^3, and in like manner the tumbling shaft arm will determine its centre of shaft T.*

V. To find the lead of the forward and return strokes in the full gear.

Having swept with the hanger an arc c^2 c^2 (Fig. 36) upon which the stud travels in the full gear of the link, slide the template on the forward lead elements F, B, until its stud lies at S^7 in the full gear arc c^2 c^2, and mark the point d in which the link arc then intersects the central line of motion.

In like manner slide the template on the return stroke elements f, b, and mark the intersection d^3.

The distances of these points from the lap circle will equal their respective leads. Thus in the forward stroke the lead equals l d in the full gear, but l d' in the mid gear ; while l' d^3 equals the lead opening of the full gear return stroke, but l' d^2 of its mid gear. In the present case both mid-gear leads were made equal to each other. Their slight variation in the full gear has absolutely no effect on the motion.

VI. Extreme travel and slip of the link.

Referring to Fig. 34 we observe that the forward eccentric attains the extreme points of its throw at D and E on the central line of motion at which times the backing eccentric occupies the positions T and U. [The latter points may be laid off from D and E with a pair of dividers set to the distance F b.] By sweeping the elementary arcs of the eccentric rod pins for these points and adjusting the template thereto, we obtain the positions Nos. 9 and 10

---

* If the ratio of crank to connecting rod had been that of 1 : 5 or 6, the lines c^2 c^2, c^3 c^3 would have had a greater inclination to the central line of motion, thereby removing h and h^3 to h^4 and h^5, and depressing the shaft to some impracticable point T^2, where it would have been brought in contact with the forward eccentric rod when the link was in the back gear. The proper adjustment for such a case will shortly receive our attention.

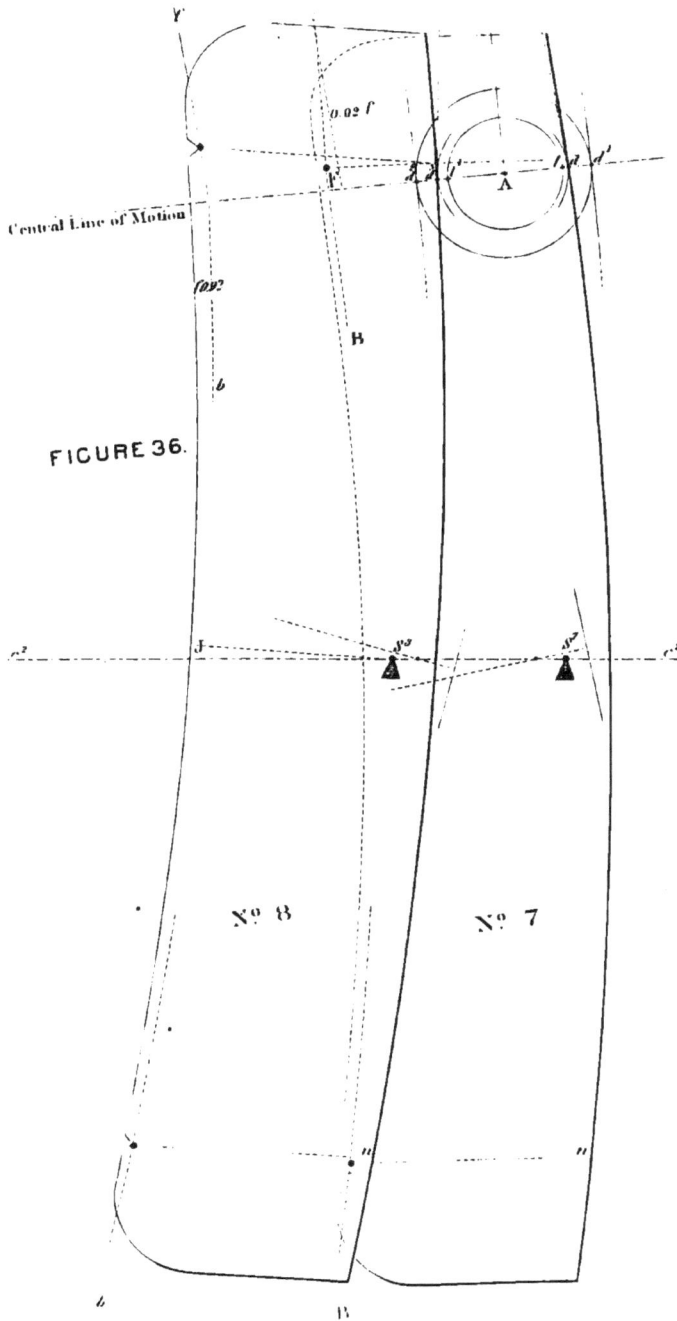

FIGURE 36.

Central Line of Motion

No 8

No 7

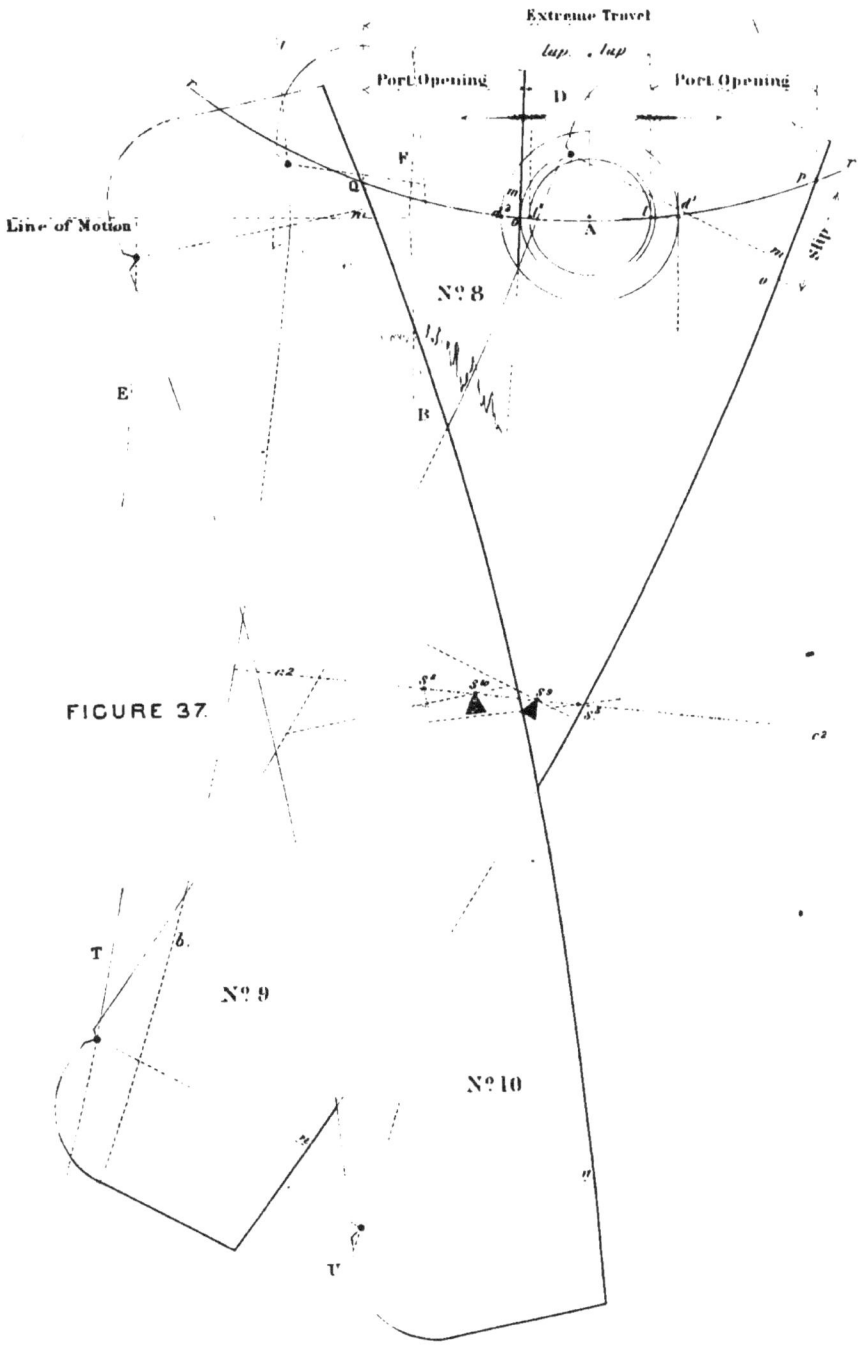

Extreme Travel

lap. lap

Port Opening   D   Port Opening

Line of Motion

N.° 8

F

E

R

FIGURE 37

No 9

T

No 10

U

(Fig. 37) and are able to mark the extreme points Q. p. of the rocker arc which are separated by a horizontal distance equal to the extreme travel.

For position No. 8 the fixed point m on the template attains its maximum elevation above the link arc, and now, at the extreme throw, its greatest depression below that arc. The maximum slip will consequently equal the distance from p to o on the link arc. This slip grows less and less the *nearer* the stud approaches the rocker pin, and if the intention should be to use the link principally in the $\frac{1}{2}$ gear this amount of slip would not prove detrimental.

MODIFICATIONS.

We have thus far, as concisely as possible, presented a geometric method for determining the proportions of all link motions, similar to those illustrated in Fig. 22. But its application cannot be considered *universal*, until certain expedients are explained, by which some of the results may be varied at will and also the motion corrected, when the ratio of crank to connecting rod is other than that of $1:7\frac{1}{2}$.

I. How to reduce the slip.

In the link motion of Fig. 37 the greatest slip occurs in the full gear and the least in the mid gear. Now it frequently happens that after designing a motion the maximum slip is too great for practical purposes and the query arises: what change should be made for effecting its reduction?

There are four varieties of alterations capable of accomplishing this object, which we will here mention in the order of their relative efficiency.

1st. *Increase the Angular Advance.*
2d. *Reduce the Travel.*
3d. *Increase the Length of the Link.*
4th. *Shorten the Eccentric Rods.*

Any one or more of these agencies may be employed at the discretion of the designer and a more perfect motion be produced. Thus we could diminish the slip to ⅓ of its present value, by either increasing the angular advance to 30°, or by reducing the travel to 4 inches. Of course any such change involves an entire reconstruction of the motion in accordance with the principles already explained.

II. How the slip may be distributed.

Referring to Fig. 24 we observe that the general tendency of the fixed point m is to move in an arc the *reverse* of that pertaining to the rocker pin, while n traverses one more or less parallel. The maximum slip of the forward gear consequently exceeds that of the back gear. These quantities can be equalized in great measure by placing the stud, not on the central line j S between f and b but, upon some parallel line nearer to f than to b, usually from 2 to 2½ inches above j S. In such case the proper location for the stud is found by first drawing such a suspension line upon the link template, then sliding the template to the positions Nos. 3, 4, 5 and 6 for the forward motion, and locating the line in question at each of these positions. Make similar locations (found with the proper arcs) for the back motion. Finally locate the stud for the full gear forward in such a manner that the line c^2 c^2 joining its points shall be parallel to the central line of motion. Two or three trials may be necessary before a suitable height for the suspension line above j S is obtained. This mode of suspension is frequently adopted on locomotive link motions.

On the other hand, where no importance is attached to

the accuracy of the back motion, the slip may be greatly diminished by inclining the rocker arms to each other (Fig. 28) so that the arc r r of the pin shall, instead of preserving a state of tangency to the central line of motion, intersect it in a similar manner to the path of the point m, Fig. 24. This method can be employed with advantage in designing marine link motions.

III. SHORT CONNECTING RODS.

If the ratio of crank to connecting rod had been assumed at $1 : 4\frac{1}{2}$ instead of $7\frac{1}{2}$, the position of the stud S found as before for equalized cut-off at the $\frac{1}{2}$ stroke, and the template slid to the positions 5 and 6, the change would have impressed on the line c^2 c^2 an inclination of $18°$ (instead of $5°$) to the central line of motion. This would have rendered it impossible to equalize the motion in the ordinary way for more than *one direction* of the crank, without bringing the tumbling shaft and eccentric rods in conflict. But if the aim be simply to equalize the forward without regard to the back motion—*a very common practice*—no special difficulty will be experienced in hanging the tumbling shaft to even this case, for the point h^3, Fig. 35, would then be left out of the account.

There exists, however, a method of equalization which corrects the difficulty for both the forward and the back gears. It is clear that the tumbling shaft is employed most conveniently and successfully, when it sustains the hanger in such a manner as to *guide its vibrations in arcs practically parallel to the central line of motion.* Hence if we wish the link to conform with this condition it will be necessary to raise the template from position No. 6 of Fig. 34 to No. 11 of Fig. 38 in which the line c^2 c^2 becomes *parallel* to the central line of motion. This elevation moves the lap point l' to l^2 giving thereby a smaller lap circle l l^2 from which to determine the lap, and at the same time increasing the lead opening of the return stroke. The position A of the rocker is thereby carried to a point a more remote from the shaft, and the lead opening of the return stroke in the mid gear becomes greater than that for the forward gear. But we have already seen that whatever inequalities of lead opening may arise in the full gear, none can be tolerated in the mid gear. Nor is there an occasion for their existence, because the *link arc* may be struck with a SHORTER *radius* than the distance from the shaft to the rocker and all such inequalities be entirely eliminated.

Take in a pair of compasses. the radius A d' of the mid-gear travel and strike the circle d^4 d^5 about the new position a of the rocker. To this the link arc must be *tangent* when the template is placed at the mid-gear position No. 1 and 2. Bring the template to position No. 2, mark the fixed points m and n of the full gears on the paper, and then search for a radius, having its centre in the central line of motion, whose arc shall embrace the three points m, d^5, n.

In the present extreme case the radius equals $41\frac{1}{2}''$ against $49\frac{1}{4}''$ employed with the ratio of $1 : 7\frac{1}{2}$. The two little cuts V and Z, Fig. 39, illustrate the character of the lead openings before and after the change of the link arc radius.

Having thus determined the *true radius* for the link arc a *new* template should be constructed, and all the locations made which are appropriate to the template positions Nos. 1, 2, 3, 4, 5, 6, 7, 8, 9, 10, under the new conditions, just as though no previous investigation had taken place.

The result will be a motion capable of ready suspension from a tumbling shaft, with perfectly equalized cut-offs, with port openings varying to a slight extent in the full

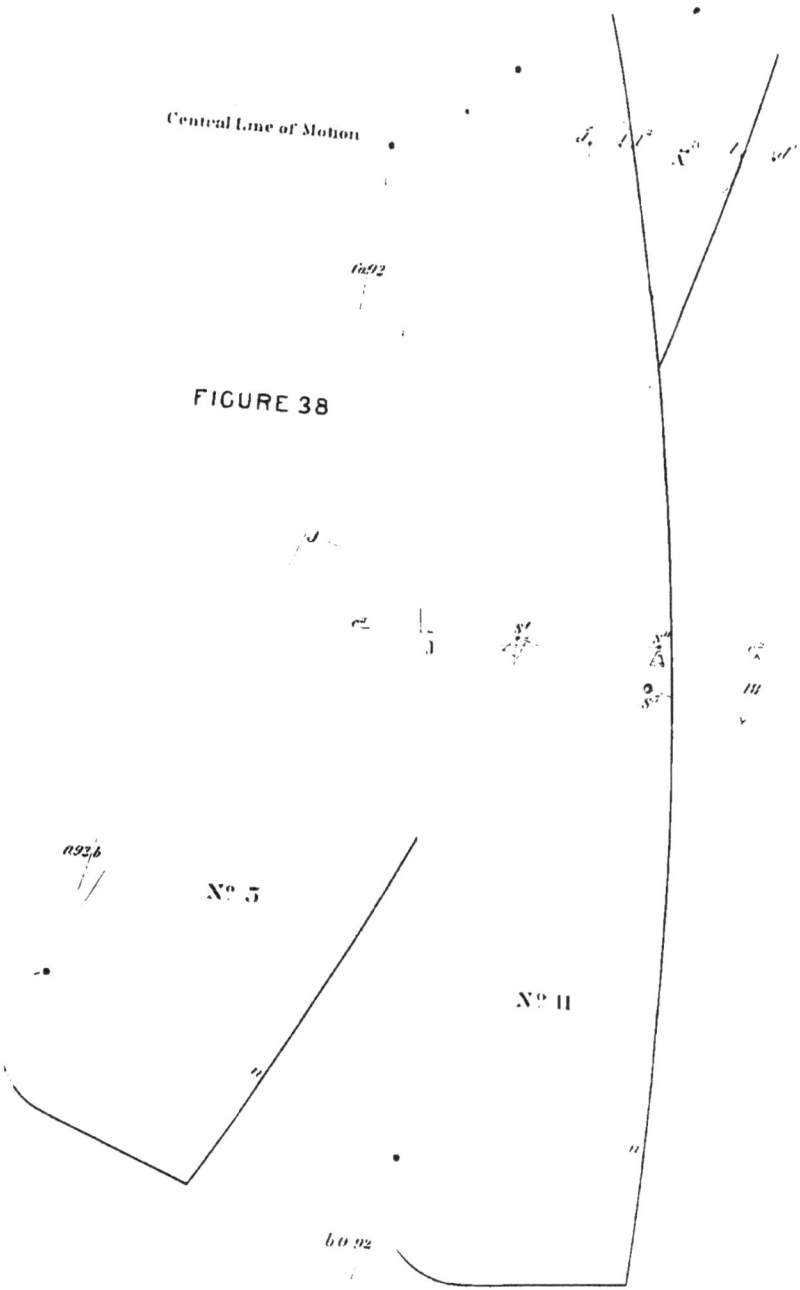

a92c

Central Line of Motion

b92

FIGURE 38

No. 5

a92,b

No. 11

b 0 92

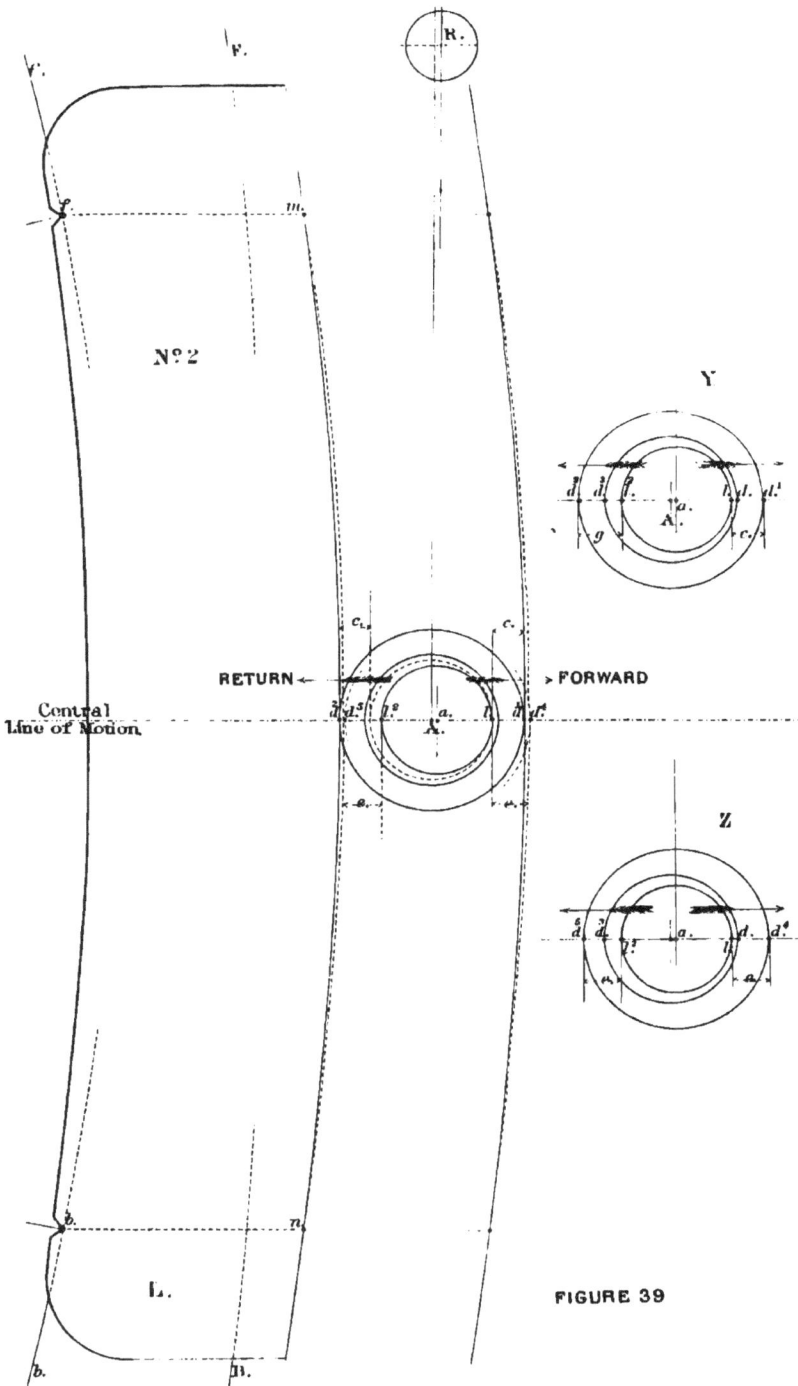

FIGURE 39

gears but precisely equal in the mid gear, with a forward
stroke lead opening increasing from l d (Fig. Z, in the full
to l d^4 in the mid gear, and with a return stroke lead, opening gradually from l^2 d^3 in the full to l^2 d^5 in the mid gear.
Both these lead openings will be *exactly* equal to each other
in the latter gear.   It will be observed that all the inequalities of the motion are thus brought in the full gears, the
very position where their influence is the least injurious.

Of course it is not pretended that great inequalities of
port openings are admissible in the full gears, unless the
smallest of these openings shall exceed the 0.6 or 0.9 area
(mentioned on page 23), in which case the magnitude of
their difference becomes a matter of little importance, because the wide opening will then admit no more steam than
the narrow.   But since the 0.6 or 0.9 area must be reached
before the mid gear is attained, where the areas become
equal, it is desirable to have as little irregularity in the
other openings as possible.

In reference to the lead opening, it is not unusual in locomotive "stationary-link" motions to give a *constant*
lead of $\frac{3}{8}$ inch, where one increasing from $\frac{1}{16}''$ or an $\frac{1}{3}''$ in the
full gear to $\frac{3}{8}''$ in the mid, would be employed under like
circumstances for a shifting link motion, and so far as can
be discovered both motions are productive of equally good
results.   Hence we fail to perceive the force of objections
so frequently urged against a slightly irregular lead, but
believe that within proper limits resort should be had to
this *most efficient means* for correcting the inequalities due
to a short connecting rod.   The recommendation, however,
must be qualified for marine engines, whose more massive
reciprocating parts require *equal* admissions for bringing
them smoothly to a state of rest at the extremities of the
stroke.   For all such it is better to equalize the lead open-

ings, at least in the full gear, and as far as practicable the
cut-offs for the same, paying but little attention to the back
gear.

It has doubtless been observed that throughout the pre
vious investigation no allusion has been made to the equality of the exhaust closure, and no direct effort put forth for
its conservation. The omission was intentional, simply because the very process of equalizing the cut-off incidentally
accomplished this object. In proof of which, it is only necessary to consider that the neutral position (A) or exhaust-
closure point lies *exactly midway* between the cut-off points
l, l'. so that if the motion be corrected for the latter when
these are near each other, it must become practically perfect for the former. But if the maximum cut-off should
take place at about the ¾ stroke of the piston, the lap points
l and l' would be widely separated, and probably give rise
to a marked inequality of the exhaust closure at the mid
gear. Since irregularities of closure and release produce a
greater impression on the motion when they occur at the
mid rather than at the full gear of the link, and since it is
possible by means of inside lap and clearance to correct
them for either of these gears, it will be well to determine
the extent of the inequality for the *mid gear*, and regulate
accordingly the position of the exhaust chamber with reference to the edges of the valve. To determine this correction, bring the link template to one set of the half-stroke
elements (Fig. 32) and slide it thereon until the link arc
stands over the exhaust-closure point A (or a, as the case
may be). Mark the position of the stud S, and through
this point draw a line parallel to the central line of motion.
Next move the template on the other set of ½ stroke elements until the stud reaches the line just determined.
Mark the point in which the link arc now intersects the cen-

tral line of motion, and measure the distance of such intersection from A. The required correction will be ½ *of this quantity.* If the point falls on the forward side of the valve stroke it indicates that the exhaust chamber must be moved *bodily* this amount towards the forward edge F (Fig. 11); but if on the back. towards the back edge N.

IV. ECCENTRIC ROD PINS BACK OF LINK ARC.

Judging from the frequency with which mistakes are made in the location of the eccentric rod pins, one is apt to conclude that most Designers regard their arrangement as a mere matter of caprice, and as having little or no bearing on the symmetry of the motion. This, however, is by no. means the case, for each combination has its own appropriate method of attachment. A connection of the pins *back* of the link arc is best suited to a motion like that of Fig. 22, because it tends to *hasten* the speed of the rocker pin in the forward stroke. an object usually accomplished by raising the link at the expense of slip.

The upper part of Fig. 34 has been reproduced in Fig. 40 for the purpose of explaining this peculiarity of Link No. 1. By erecting perpendiculars to the central line of motion through the cut-off points *l, l',* and by measuring from the same line, the respective distances of the eccentric rod pin *f,* for the 0.92 cut-off elements, we discover that the forward stroke distance T is much less than the return stroke R. and that if T was equal to R the rocker pin would be carried *beyond l,* thereby *delaying* the cut-off of the forward stroke which now it *hastens.* In reality the slip is avoided by elevating the eccentric rod pin instead of the link arc. The advantage gained by this feature is very great, for while it tends to equalize the motion it reduces the slip of the link and renders easy the work of suspension from a tumbling shaft.

In locomotive practice the distance of the pins' remova¹. from the arc varies between 2¼ and 3 inches. For marine work the limits are, in general terms, double these quantities.

Fig. 40.

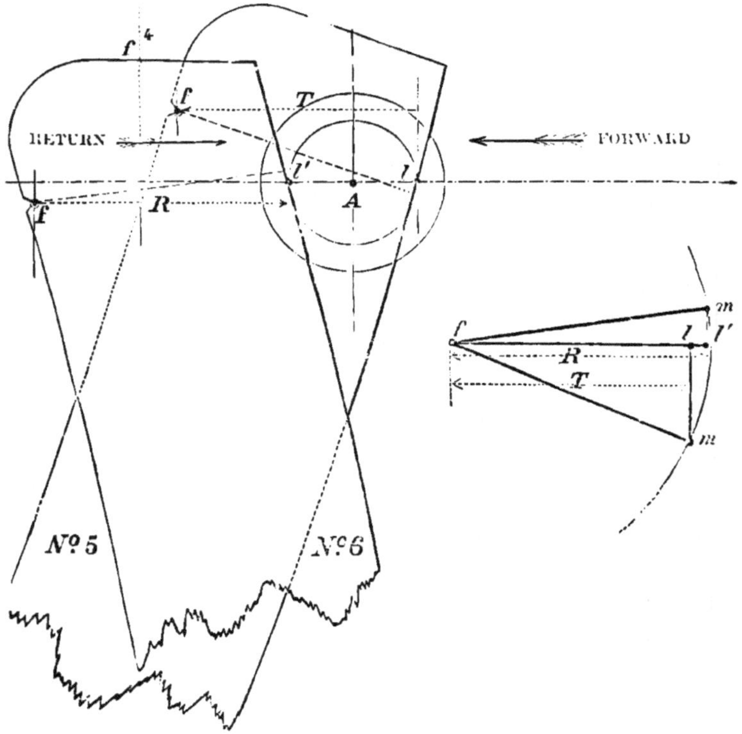

The principal fact to be borne in mind is that, within suitable limits, the *greater* this distance the *more readily* can the motion be equalized.

From the foregoing we perceive that LINK No. 1 is appropriate to a motion requiring *acceleration for any cut-off point beyond the neutral position* A, *and having its smallest crank angles laid off between the link and centre of the*

APPLICATIONS OF LINK Nº1. (Positive Motion.)

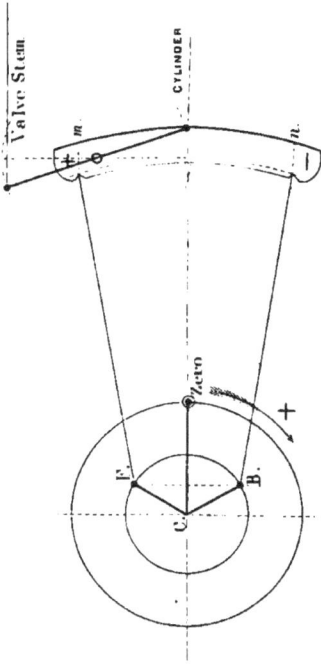

DIRECT ACTION ENGINES.

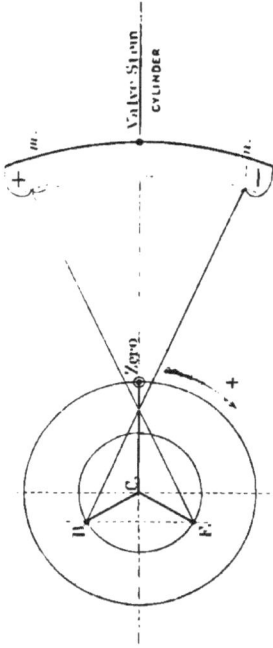

BACK ACTION ENGINES.

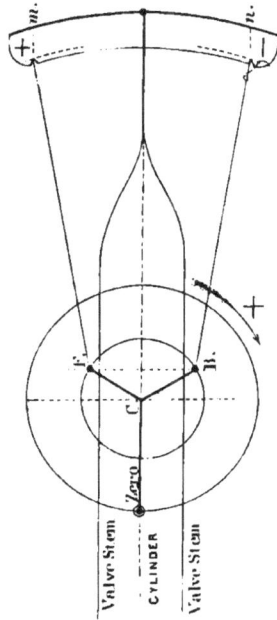

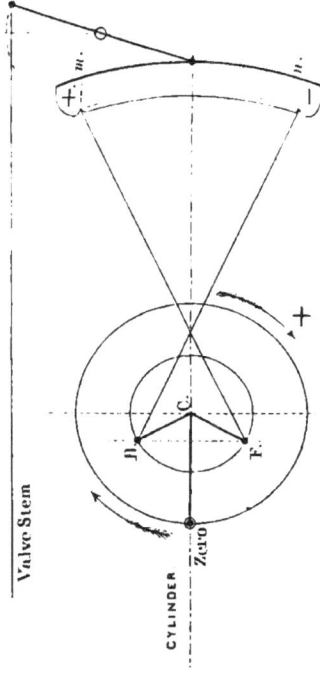

main shaft (as in Fig. 29). The four typical forms of such motions are here presented with a suitable arrangement of the parts, when the link is dropped in the full gear forward, for a positive motion of the crank. But if a negative motion be demanded, it is only necessary to transpose the entire motion so that the cylinder comes on the *opposite* side of the shaft, in other words to take the power off the opposite extremity of the shaft after first turning the engine end for end.

It was remarked (page 61) that the crank angles of a *back-action* engine are invariably the *reverse* of those common to one of direct action; hence, if we wish to preserve the identity of the motion in laying off a figure like 29 for such an engine, it will be necessary to locate the initial positions F and B of the eccentrics opposite to those proper for a direct action, and apply thereto either of the two schemes offered in the Diagram.

EXAMPLE.

The following dimensions have been taken from a most successful freight engine; their application will serve to familiarize the student with the principles of the foregoing method: 6 driving wheels, 57 inches diameter. Outside cylinders, 18 inches × 22 inches. Connecting rods 86 inches in length. Ports 15″ in length; steam, 1¾″ wide; exhaust, 3″. Diameter of eccentric circle=5 inches. Maximum cut-off=0.8 of stroke. Mid gear lead=$\frac{5}{16}$ inch. Rocker from shaft=55¼ inches. Length of rocker arms=9 inches. c. to c. of eccentric pins=13 inches. Tumbling shaft arm= 18 inches. Hanger=13¼ inches. Pins back of arc=3¼ inches.

Required.—Radius of link, distance of point of suspension back of link arc. lap of valve, full gear lead, and location of the tumbling shaft.

COMMONLY known as the "Open Link," is specially adapted to cases in which the link acts directly on the valve stem without the intervention of a rocker, as peculiar to British practice. It differs from No. 1 in the location of its eccentric rod pins. These, instead of occupying stations back of the link arc, reside at points f and b beyond the extreme positions m and n of the link block. They consequently move a greater distance than the latter points, and in order to preserve the same travel of valve the eccentric circle must be enlarged.

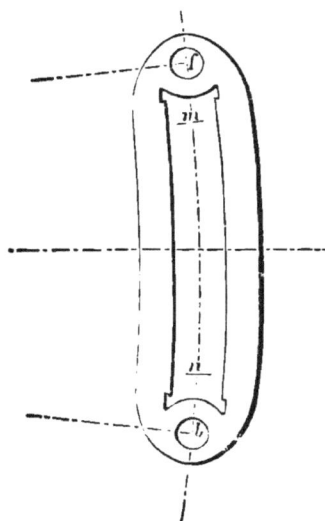

In locomotive practice the distance between eccentric rod pins f and b varies from 16″ to 20″.

The diameter of the eccentric circle varies from $5\frac{1}{2}$ to 7 inches, and the working points m and n usually about 3″ from the pins. The template for this form of link is illustrated by Fig. 41.

This link No. 2 is specially adapted to a motion requiring acceleration for any cut-off point beyond the neutral position A, and having its *greatest* crank angles laid off between the link and centre of the main shaft. It will be remembered that with Link No. 1 (Fig. 40) the *smallest* crank angles occupied such a station, and the eccentric pin was elevated to the position f, making the distance T less than R. If, however, the *greatest* crank angles had occupied this position, the element f would have been removed

FIG. 41.

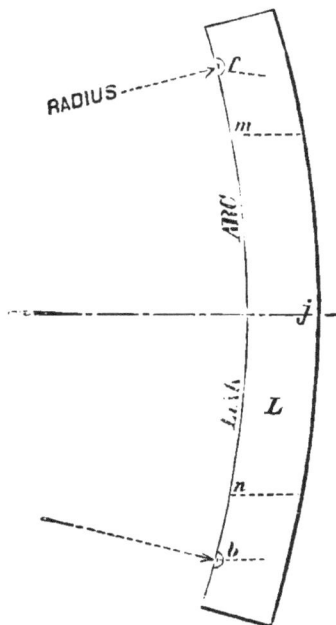

to some line f^4 nearer the shaft, while the link would have been depressed and slip increased in the endeavor to make T=R. But the equality of these terms is established when the link arc passes through the centres of the eccentric rod pins: hence, a "Box Link," having its pins over the working points m and n, will be found better adapted to this position than Link No. 1. yet it only supplants the more readily-constructed open link, when it becomes important to retain the throw of the eccentric at its lowest limit.

As with Link No. 1, so with No. 2, there are four

9

schemes to which the motion is peculiarly applicable. These are given in the accompanying Diagram for the positive motion of direct and back-action Engines. As heretofore explained, their negative motion may be obtained by transposing the cylinder and link motion to the opposite side of the main shaft and taking off the power from the other extremity of the same. When, therefore, it has been decided that an engine shall have a direct or back-action arrangement of the connecting rod, the link to act through or without a rocker on its valve, and its forward motion to be positive or negative, then an appropriate arrangement of the link can be at once selected from these two diagrams, and the side of the shaft determined, on which the cylinder should be placed, to secure the desired crank motion when the link is dropped in the full gear forward.

It should be observed here, that the lower part of the link is never used in a horizontal engine to impart a forward motion because the weight of the overhanging or sustained parts tends to render the motion unsteady.

CONSTRUCTION

To find approximately the throw of the eccentrics and their angular advance for a given travel of the valve, length of link and position of the working point.

Describe about any point C (Fig. 42) on a vertical line F R a semi-circle E H with a diameter equal to the proposed travel of the valve; from C lay off C P, the extreme distance of the working point from the eccentric pin, also C S the ½ length of the link. With S as a centre describe the arcs C f, P m, and b R. Suppose now the cut-off must take place at 0.78 of the stroke, then from the Travel Scale

we learn that the trial angular advance should be 28 . Lay
off C B 28 from F C and produce the line D B passing

FIG. 42.

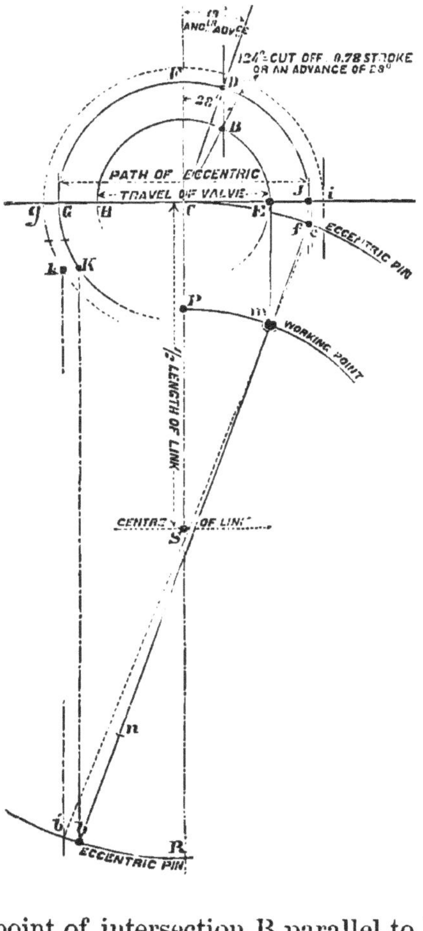

through the point of intersection B parallel to F R. Strike
a trial eccentric circle j F K about the centre C. Take the
distance between its two intersections D and F in a pair of
dividers and lay it off *twice* from G, giving the point K.

Through K draw a line parallel to C R and determine its intersection *b* with the arc R *b*. Draw a right line through *b* and the extreme travel point *m*. If now *f* should be found in the tangent line through *j* to the eccentric circle, it would prove that the diameter of this circle had been assumed correctly. But if it falls *without* the assumed circle, this diameter must be *increased*. Conversely any point *within* requires a *diminution* of the same. Thus the point *e* indicates that the diameter of its circle *j g k* has been assumed too large.

Having secured the correct diameter of the eccentric circle, join the points D and C by a right line and measure its inclination to the line F R. We thus obtain 19° the true angular advance of the eccentric for accomplishing the desired cut-off. The principle involved in this construction is that when one eccentric produces its extreme throw D (Fig. 33) the other will be separated from its like position E by the horizontal distance of T, which always equals *double* the angular advance. Although this construction does not claim strict accuracy, it will be found to answer all practical purposes.

APPLICATION OF LINK NO. II.

For illustrating the process of manipulation peculiar to this form of link, we will assume the following terms :

Crank and rod ratio $=1 : 6\frac{1}{2}$.

Cut-off at 0.8 of the stroke.

Travel of valve $=4\frac{3}{4}$ inches.

Valve stem from shaft $=5$ ft.

Distance between eccentric rod pins $=18''$.

Extreme working points, $3''$ from pins.

Mid-gear lead opening $=\frac{3}{8}$ inch.

APPLICATIONS OF LINK Nº 2. (Positive Motion.)

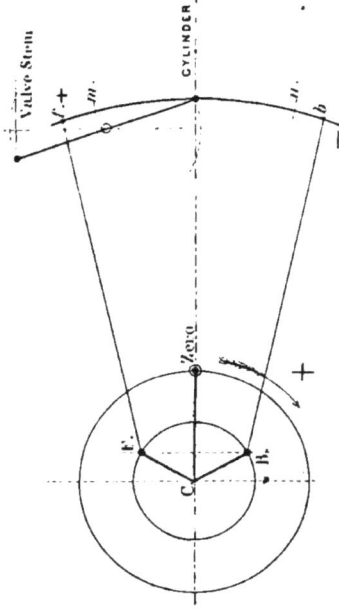

BACK ACTION ENGINES

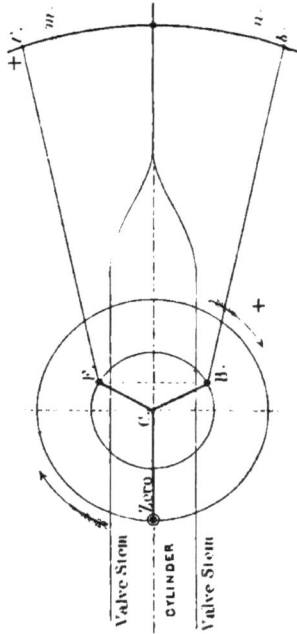

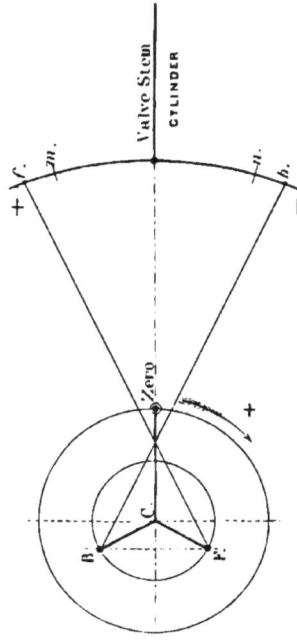

"DIRECT ACTION ENGINES"

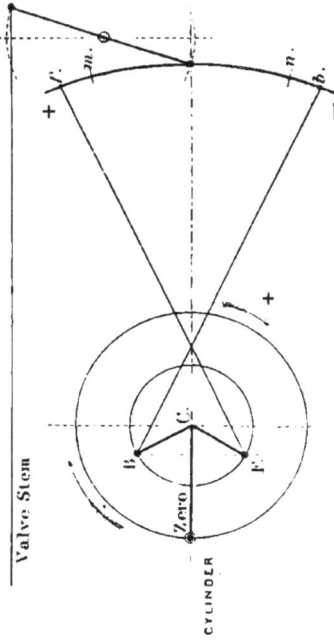

By a construction like that of Fig. 42, an angular ad
vance of about 16° and eccentric circle of 6¾ inches diameter
are found to conform with the above conditions. If now
the connecting rod and link motion act *directly* on the piston
and valve the initial position of the motion will be that presented in the first Figure of the APPLICATIONS, and the
four starting points F. B. *f*, *b*, together with the four ½ stroke
points of the eccentric circle, may be mapped in a similar
manner to Fig. 29 ; while the four 0.8 cut-off points should
be laid off by means of a protractor from the lines F C, B
C, *f* C, and *b* C.

Having mapped these twelve important positions and
constructed a template like Fig. 41 the next step will be to
follow the journeyings of the link arc and locate the tumbling shaft in a manner precisely analogous to that pursued
with Link No. 1 :

Find
{ 1st. The mid-gear travel $=d'\ d^2$.
2d. The lap A *l* for a given mid-gear lead.
3d. The position of the stud for equal ½ stroke
cut-offs of the piston.

This form of link is more generally suspended from the
upper eccentric rod pin than from a point midway between
the two, and has the tumbling shaft *below* the central line
of motion ; but by marking the various locations of both
pins, as well as the centre line *j*, we will be able to select
the most appropriate of the three suspensions.

Let us follow first the central suspension. We find that
by locating the stud on the arc, the line *c c* of its motion,
for the ½ stroke cut-off, becomes parallel to the central line
of motion, a condition suited to ready suspension. But on

dropping the template to the full gear cut-offs the stud assumes the positions S^4 S^5, Fig. 43, whose line c^2 c^2 has sc
great an inclination to the central line of motion, that it
would be next to impossible to successfully hang it from a
tumbling shaft and accomplish *equal* cut-offs in *all* gears.
We must consequently resort to a change of link arc radius
and unequal full gear leads (as in Figs. 38 and 39) in order
to establish perfect equality. To do this, raise the link arc
so that the stud shall occupy a location S^{0*} thereby bringing
c^2 c^2 more nearly to a state of parallelism with the central
line of motion.   Mark l^2 the new lap point ; strike a new
lap circle l l^2 with neutral position a, *nearer* to the main
shaft, and about this point describe the mid-gear travel circle giving new points d^4 d^5 through which the link arc must
pass for the mid gear.   Bring the template to position No.
2, as in Fig. 31, mark the eccentric pin points f and b ; then
search for a radius whose arc shall pass through the three
points f, d^5, b.   This radius will always prove *greater* than
the distance from the centre of the shaft to the neutral position A instead of *less*, as found with Link No. 1.

Having obtained a correct link arc, cut out a new template like Fig. 41 to represent it, and reconstruct the entire
motion.   The new radius will equal 5 ft. 7″.   The corrected
lap $= 1\frac{1}{32}$ inches.   The lead for the forward stroke, increases
from $\frac{1}{8}$ to $\frac{3}{8}$ in the mid gear, and from $\frac{3}{16}$ to $\frac{3}{8}$ for the return
stroke.   The lines of suspension pin motion c, c', c^2, c^3 may
converge, as in the present instance, towards some point
beyond the link.   In such case the tumbling shaft should
be placed on the side of the convergence.

If, however, the link is suspended by the upper eccentric rod pin, the forward gear paths of the suspension pin
motion will be represented by the lines h, h', and the tumbling shaft will lie below the central line of motion ; but for

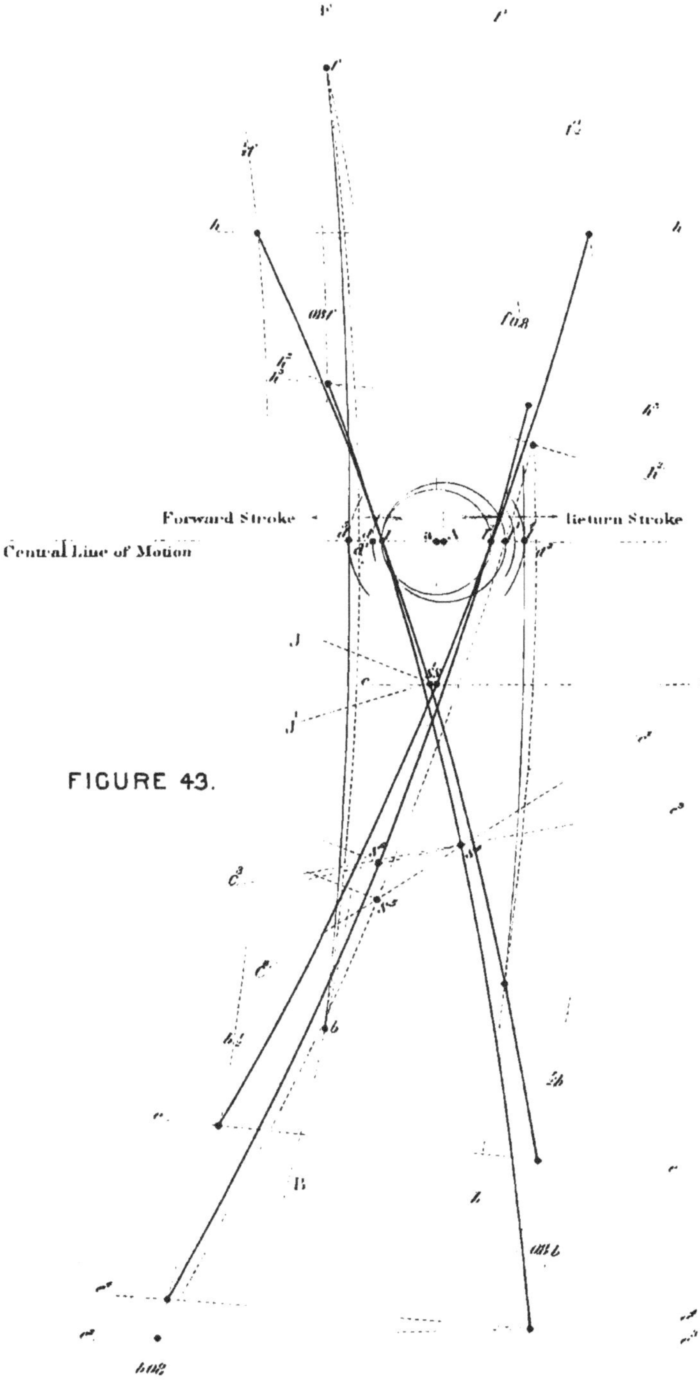

Forward Stroke ←

Central Line of Motion

→ Return Stroke

FIGURE 43.

the lower eccentric rod pin, the lines will become c, c', c', and its tumbling shaft will stand above this line.

The judgment of the designer will in every case decide to what extent the lead openings should vary in the full gears, what inequalities of cut-off are admissible, and whether or not the motion should be equalized simply for the forward without regard to the back gear.

While attempting to analyze the peculiarities of various existing link motions, the investigator usually finds that among other dimensions, only the lap and mid gear lead are given, and that no allusion is made to the angular advance, or to the maximum cut-off. To solve such cases, he is compelled to work the problem as it were, *backwards*. On the central line of motion he plots the lap and lead, as on Fig. 32, places the template in the Nos. 1 and 2 positions, as in Fig. 31, then with the length of the eccentric rod as a radius and the positions of the eccentric pins as centres, strikes in the direction of the shaft indefinite arcs, whose intersections with the eccentric circle give the points F, B, f, b, (equi-distant from the central line of motion) and from these the desired angular advance can readily be measured.

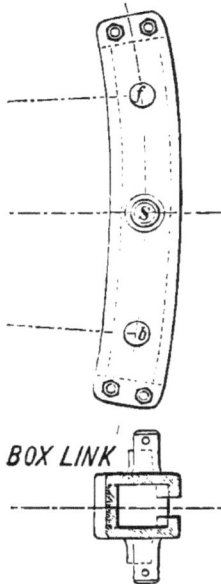

BOX LINK

ALTHOUGH the mechanical construction of this form of link is rather more difficult than that of No. 2, it serves the part of a good substitute when a short throw of the eccentric is required ; for with it the maximum travel of the valve always approximately equals the diameter of the eccentric circle.

At times the box link can be employed in positions appropriate to Link No. 1, but *very rarely* with those *good results* in respect to minimum slip which obtain with the former motion. On such occasions the stud usually lies at some point *beyond* the link arc, determined by placing the link in the ¼ stroke cut-off positions, Nos. 3 and 4, and plotting the centre line *j* as in Fig. 32.

When, however, the box link is used in place of Link No. 2, the centre of suspension S generally falls *within* the link arc, or between it and the main shaft.

The construction of these links varies ; in some instances the ribs are formed on the inside as represented, while in others, they are cast on the sliding block and overlap the link plates.

THIS form of connection between the valve and eccentrics is specially applicable to those circumstances in which the former requires no rocker. The mutual relation of the parts will be clearly perceived from an examination of Fig. 44 which illustrates one of the most successful methods of suspension.* The eccentrics stand in their usual location for a direct action motion. The main link is hung from a *fixed* point by a short bar called the "suspending link" and the link block connected with the valve stem through the " Radius Rod " $m\, d'$. By means of a reversing combination the block may be carried to any point between m the full gear forward and n the full gear back. But since the link arc is always struck with a radius equal to the length of the rod $m\, d'$, having its centre at d' and d^2 in the central line of motion, when the crank occupies the zero or $180°$ location, it must be evident that the block can be moved from one full gear to the other *without* altering the position of the points d' or d^2, consequently the lead opening will remain *constant* throughout the motion. . . . . Now it has been invariably the custom to simply define a stationary link motion as "*one in which the lead is constant*"

* For convenience of observation, the cross sections of the valve and seat have been revolved to a plane at right angles to their true position.

leaving it to be inferred that the angular withdrawal of the crank from its zero position at the moment of pre-admission must also be a *constant* quantity, whereas in reality this lead angle *increases* just as much for a stationary link motion as for a shifting one. The only difference between the two is that the lead opening of the stationary link motion is more ample and the angle slightly greater, for all except the mid gear, than with the shifting link motion. But this distinction has been so clearly shown heretofore that further remark can scarcely be necessary. Unlike the shifting link motion, however, the lead opening is *not* dependent on the arrangement of the eccentric rods, for these may either be crossed or opened without altering the result. But for the purpose of meeting the other conditions of the motion an arrangement like Fig. 44 should be adopted.

As a general thing more attention is paid to the equalization of the cut-off and reduction of the slip in the forward than in the back gear. For the accomplishment of this object, the centre of the link should be dropped *below* the central line of motion, the angular advance of the backing eccentric slightly reduced and the backing eccentric rod lengthened.

The simplest method by which the student can obtain a clear idea of the action of the parts, in a stationary link motion, will be for him to take the dimensions of some successful motion, cut out a proper template for the link and trace its journeyings throughout the different gears in conformity with principles already laid down for the shifting link motion.

The following dimensions (in absence of others) will answer such a purpose :

Diameter of piston=18 inches.

Stroke=24"; Connecting rod=91".

Ratio=1 : 7½. Throw of eccentrics=2⅝".

Forward eccentric angular adv. =27½°. Rod=57¾".

Backing eccentric angular adv.=26°. Rod=58".

Eccentric rod pins 12½" apart, 3" back of arc.

Centre of suspension 1½" back of arc, 1⅛" below line.

Radius rod=37", Reversing link=11½", Hanger=9".

Tumbling shaft arm=18". Reversing pin 8" back of arc

Lead=⅜", Steam port =2", Exhaust=3½".

Maximum travel about 4¾ inches.

The Stationary Link is seldom found in American practice from the fact that all modern locomotives are built with steam chests on top of their cylinders, instead of at the side. On stationary engines, the link and governor are occasionally used conjointly ; in such instances the stationary link will be found best adapted to the requirements of the case, because its radius rod imposes a far lighter duty upon the balls of the governor, than the shifting link with its rods, hanger, and additional friction of eccentric straps.

THE discovery of this motion was a natural sequence to the invention of the shifting and stationary links. By it a compromise has been effected between the leading features of both motions, resulting in a more *direct* action and perfect balance of the parts together with a reduced slip of the link block. One mode of suspension, for the link in the full gear forward, appears in Fig. 45, in which the cross sections of the valve and its seat have been revolved for the purpose of more plainly exhibiting their relative positions. The locations of the point of suspension and attachments of the eccentric rod pins upon or back of the link arc are quite as variable for this, as for the shifting link motion; and the requirements of the other details generally indicate whether the reversing shaft should be placed above or below the central line of motion.

In proportioning the parts, the main object is to move the link and radius rod (when the crank stands at the zero or 180° locations) in such a manner that the link arcs peculiar to each motion shall always be *tangent* to each other. In this case all the locations of the link block will be found in one and the same *straight* line. This peculiarity has given rise to the title "Straight Link" motion expressive of the form of the link.

The radius rod and main link are supported by rods from the reversing shaft arms, and the inequality in the lengths of the latter, which is essential to a proper suspension of the parts, incidentally tends to *equalize* the weights resting on the opposite sides of the reversing shaft, thus greatly facilitating a change of the motion from one full gear to the other.

Well-schemed motions of this type practically preserve the characteristic feature of the stationary link. viz., a constant lead; yet from the nature of the case they possess at times slight inequalities in one or both of the full gears. These, however, are quite insignificant for a relatively long radius rod and short travel.

The ratio between the long and short arms of the reversing shaft may be readily determined for any given travel, angular advance, length of eccentric rods, link and radius rod, by placing the template in the No. 1 and 2 positions (Fig. 31), marking the mid-gear travel d^{l} d^{2}, sweeping indefinite arcs through these points with the radius rod, and drawing down the template, or centre of suspension, from S to S^{l} until its straight line intersects these arcs in points m, m^{2}. Then map the radius rod m d^{l}, m^{2} d^{2} giving the points u, u^{l} of the reversing rod pin above the central line of motion. The centre l of the link arm pin must fall as much below the horizontal line through the reversing shaft R as S^{2} does below S. In like manner the pin h of the other arm must rise as far above the horizontal as u does above the central line of motion. Finally, draw l h of length sufficient to accommodate the details of the shaft, and we have at once the proper dimensions for the reversing shaft arms.

It should be observed that the tendency of m^{2} is to drop below m and thus distort the motion, but this will be obviated if the two are brought into one horizontal line *inter-*

mediate between such stations. The change will result in a *slightly* increasing lead, as is common with the shifting link motion.

The following dimensions can be employed for an investigation of this class of motions:

Diameter of cylinder=16″.
Stroke=24″ ; Connecting rod=87″; Ratio=1 : 7¼.
Throw of eccentric=2¼″ ; Angular advance=26°.
Eccentric rods=39½ inches.
Radius rod=47″, connected 7″ back of link.

Box link, suspended by stud at centre with eccentric rod pins 10″ apart.
Suspending rods both 18′ long.
Reversing lever, long arm=6″.
Reversing lever, short arm=2⅜″.
Mid-gear lead=¼ inch.
Steam port=1¼″ : Exhaust=2¾″.

———————

If the examiner desires to analyze any stationary or straight link motion already constructed, having a given lap and lead but no specified angular advance, he should work the problem *backwards*, as follows: First locate the four cardinal points *d′*, *l*, *l′*, *d²*, and sweep their arcs with the radius rod ; then place the link in the positions Nos. 1 and 2, with the positions of the eccentric rod pins as centres and the length of the eccentric rod as a radius, sweep the four arcs, which must contain the points F, B, f, b, and describe through them the given eccentric circle, in such a manner that all the points of intersection will lie equally remote from the central line of motion. With these initial points determined, the investigation can proceed on the principles already explained.

This device groups two perfectly distinct motions—the one derived from a single eccentric, the other from the cross-head of the piston rod—in such a manner that their combined effect is, when the parts are well proportioned, quite analogous to the motion obtained from the stationary link. From the nature of the connection between the cross-head and the valve-stem, the motion can be more readily applied to an outside cylinder engine than to an inside one.

The eccentric usually assumes the form of a return crank from the main crank pin, as shown in Fig. 46. Its centre then is found on a line at right angles to the crank arm. The angular advance becomes equal to zero, and hence, so far as the link will be concerned, the valve can have neither lap nor lead. The link oscillates freely about a fixed axis, and its arc has a radius equal to the length of the radius rod. This rod is moved from one full gear to the other, in the usual manner, by means of a reversing shaft with arms. From the extremity of a short arm, rigidly bolted to the cross-head pin, extends a union bar which is pinned to one end of the combination lever. By the aid of this lever, the eccentric and cross-head motions are so combined, that the latter virtually restores the angular advance discarded while locating the eccentric, and consequently enables the valve to possess both a constant lap and lead.

The truth of this assertion will appear from an examination of the lever elements (Fig. 47) for 12 locations of the crank arm.

FIG. 47.

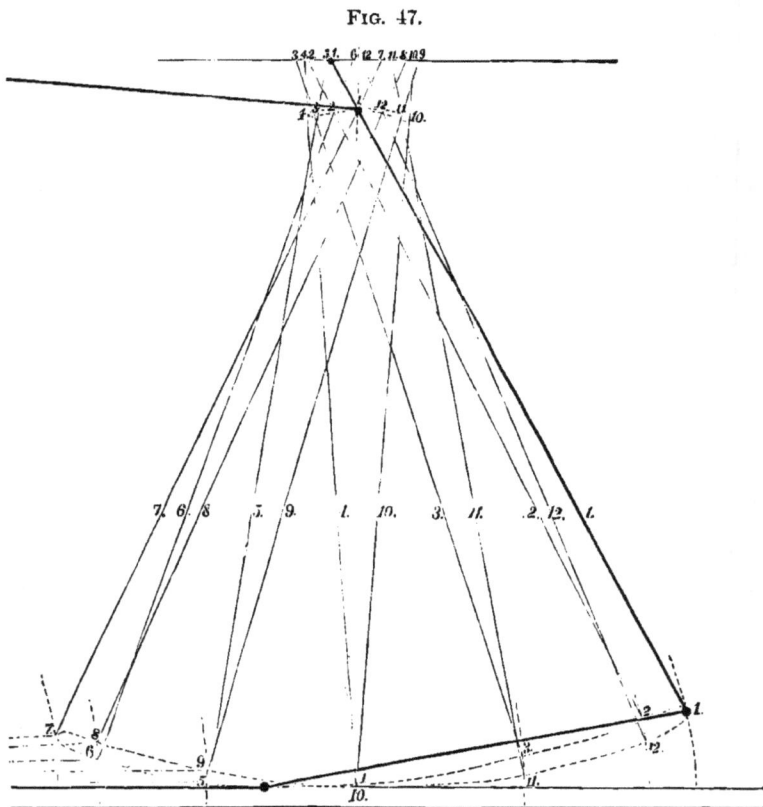

The subjoined dimensions will be found convenient for the construction of a trial example, in which of course two templates should be employed, one to represent the link, the other the combination lever :—

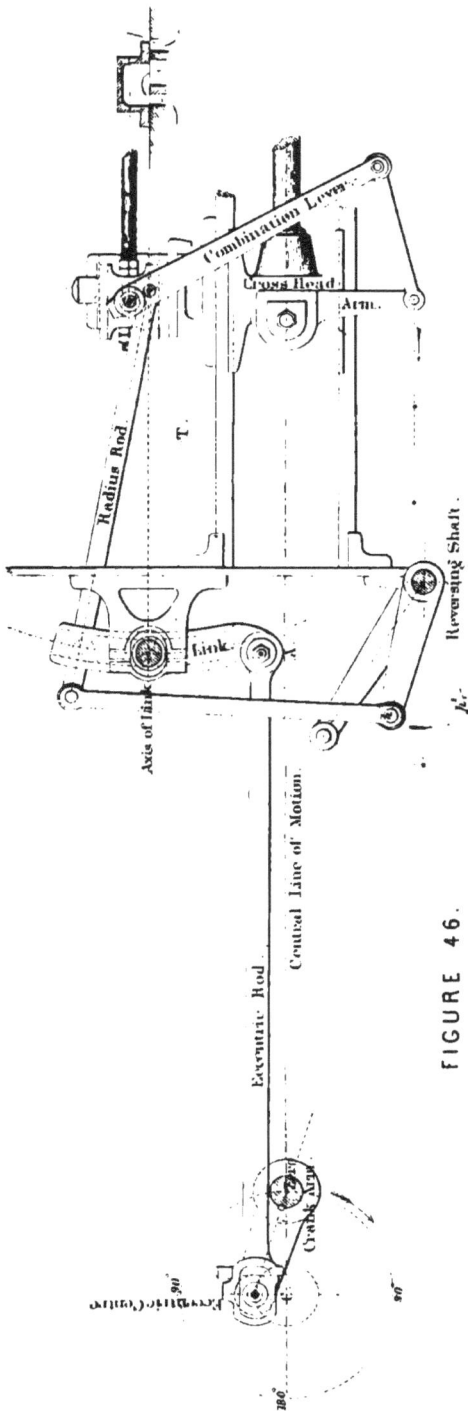

FIGURE 46.

Diameter of piston=18".

Stroke=24" ; Connecting rod=100".

Ratio=1 : 8¼.

Axis of link 74" from centre of shaft and 15¾ above central line of motion.

Eccentric rod pin 13" from axis of link.

Radius of link and length of rod=42½".

Link block to end centre of radius rod=6".

Long lever of combination arm=30".

Cross-head arm drops 14¾ inches.

Connection for arm and lever=16".

Throw of eccentric=3½ inches.

Travel of valve (maximum)=4½ inches.

Lap=1 inch ; Constant lead of ¼".

Steam port 1¼" ; Exhaust port=3".

Cases will at times arise in which the distance between the centre of shaft and valve, will prove too contracted for a satisfactory arrangement of the parts after the manner shown in Fig. 46. In such instances the curvature of the link should be reversed, the radius rod made to lie between the link and shaft, and the valve stem lengthened, to adapt it to the new position of the combination lever.

The Designer will find that his efforts, towards the equalization of the cut-off in the Walschäert Motion, are attended with far less difficulty than similar ones with the shifting and stationary links. This peculiarity arises, from the intimate relation constantly maintained between the valve and piston motions, through the medium of the combination lever. In consequence of which, any undue acceleration or retardation of the piston motion is immediately accompanied by like effect in that of the valve, thereby greatly diminishing its capacity to derange the events of the stroke.

10

PART V.

INDEPENDENT CUT-OFF,

CLEARANCE, ETC.

INDEPENDENT CUT-OFF.

IT has been demonstrated in Part I. that a very early cut-off is incompatible with the economic action of a single eccentric, for beyond the limit of about ⅔ of the stroke, the compression attending an earlier closure of the exhaust will usually furnish a resistance in *excess* of that required for neutralizing the momentum of the reciprocating parts. If therefore a more extended range of the cut-off should be demanded, for comparatively slow speed engines, other means must be sought by which it can be controlled without affecting the exhaust; in a word *independent* valves must be introduced having a motion different from that of the main valve. For this purpose the parts are usually arranged as shown in Fig. 48. The valve A carries on its back two cut-off valves B B′ and rigidly supports by pillars the surface plate H H on which a brass packing ring F F bears, enclosing a space D in communication with the condenser through the pipe E. The partial vacuum formed in this space relieves the valve in great measure of the immense pressure exerted by the steam, and consequently reduces the friction as well as facilitates the starting of the engine.

On the cut-off valve stem are turned a right and left hand thread, so that by revolving the same, the valves may be drawn closer together or separated by a wider distance according to the requirements of the cut-off. The main valve A has a lap, lead and exhaust closure appropriate to the value of the maximum cut-off, and permanently retains

FIG. 48.

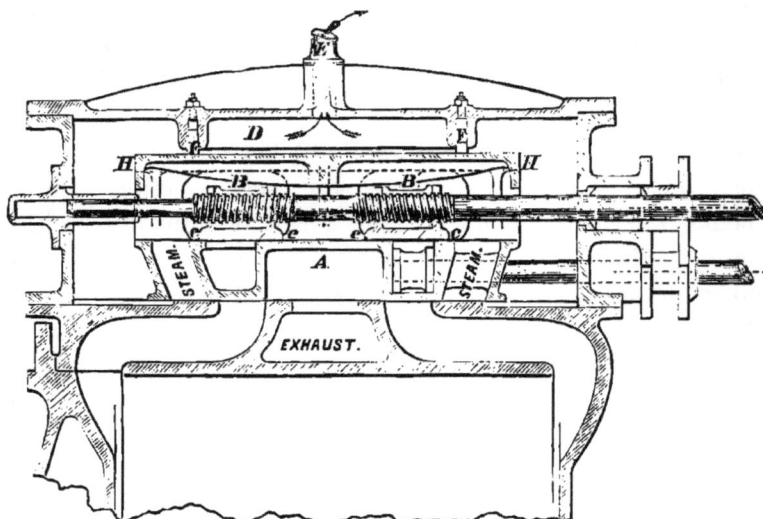

these quantities throughout every variation in the point of cut-off brought about by the separation of the valves B B'.

Its general design depends upon the range of the cut-off. For one varying between zero and about 0.6 of the stroke the combination shown in Fig. 48 is most suitable, where the *outer* edges *c c*, effect the closure of the port, and the valves have a motion directly the *reverse* of that peculiar to the piston. For cut-offs varying between 0.3 and 0.9 the stroke the *inner* edges *e e* should be made to perform this office, and the valves should have a travel *coincident* with that of the piston. Between the assigned limits such valves give sharp and decisive cut-offs for a proper ratio between the travels of the main and cut-off valves. When the *outer* edges are employed the travel of the cut-off valves should *exceed* that of the main valve ; but with the inner edges it may *equal* or exceed it according to the degree of rapidity desired in the action.

The relative motions of the valves, may for any required case be conveniently examined, by a method indicated in Fig. 49, and the travel regulated to meet the conditions of a good motion.   The valves are here supposed to cut off with their *inner* edges but the process is equally applicable to the opposite relation.   1st. Stretch a sheet of paper and upon it draw the steam ports, exhaust port, bridges, and a circle *f g* representing the motion of the main valve's eccentric with an angular advance position R 1, also the ½ stroke R 3, and the maximum cut-off R 4.   Project these points to the line of the valve seat, thus giving the positions 1, 2, 3 and 4.   Secure a slip of paper X A with its *base line* A over the point 6, and draw the main valve with lap appropriate to the maximum cut-off precisely as in Part I. Make the steam passages S S convergent in order to economize space.   Secure a second slip W D as before shown, and above its base line D describe a circle *h j* to represent the path of the eccentric acting on the cut-off valves.   Its initial position will be found at R 1 the *same* relatively as the crank of the engine.   From this lay off R 2, 3 and 4, making angles with R 1 equal to those found in the circle *f g*.   These preliminary steps completed, our object will be to find the degree of separation required between the valves for the given cut-offs, their width, and also to what extent the faces L L should be lengthened to prevent a fall of either cut-off valve.   Suppose for instance the limits were ½ and ⅞ stroke, we *first* loosen the two strips W and X, place their base lines A and D at Nos. 3 the ½ stroke locations for their respective eccentrics, and mark e for the point at which the cut-off valve should stand to effect the closure of the steam port. *Second*, move the slips W and X until their base lines A and D correspond with Nos. 4, the other cut-off limit mark e'' and it will appear that the valve stem must be rotated, until the cut-off valve B moves a distance n from its first position e. Since the right and left hand threads have one pitch, the other cut-off valve B' will be moved a like distance from the common centre C. *Third*, the valves must be sufficiently wide to guard against a *re-opening* of the port before its final closure by the main valve. This distance may be found by placing the strips at the Nos. 4 positions and marking a point c on the strip W opposite the edge d; when the length of the valve should at least equal the distance from e to c. *Fourth*, Place the strips W and X at the Nos. 1 locations and we obtain substantially the extreme position of the cut-off valve with reference to the main valve which will indicate the proper extension for the faces L L.

If other positions of the eccentric are interpolated between 1, 2, 3, and 4, the relative motions may be accurately traced and the degree of port opening observed, so that in event of the latter proving inadequate, the diameter of the circle $h j$ can be increased a suitable amount. It is customary in proportioning stationary engines with valves constructed on this principle, to make the cut-off variable between 0.3 and ¾ the stroke, while the main valve is arranged with a lead angle of about 8° and a lap suitable to a cut-off of ⅝ the stroke. The resulting angular advance usually furnishes an appropriate exhaust closure. If, however, on trial the crank should not pass its centres smoothly, the angular advance of the eccentric must be increased or diminished until the proper compression is discovered for counteracting the momentum of the reciprocating parts, as

FIG. 49.

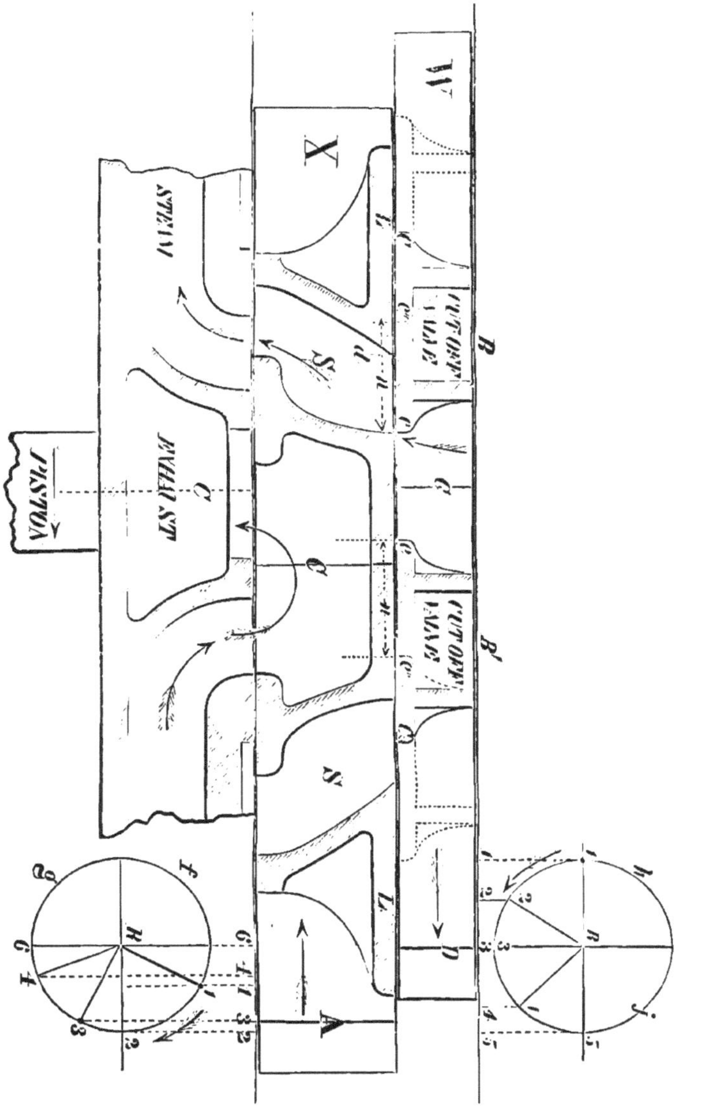

well as lead opening for neutralizing the effect of lost motion.

In marine engines the cut-off valves act with their *outer* edges. The lap angle of the main valve is then taken at from 15° to 20 degrees, and the exhaust closure effected at about 0.9 the stroke, by simply raising the link, until the eccentrics have virtually an angular advance of 35 to 45 degrees; in other words, by working the link in less than the full gear. The necessity of adjusting the eccentrics is thus obviated.

When an engine is furnished with exhaust passages perfectly distinct from those admitting the steam to the cylinder, and the demands upon its power are quite uniform, the valves regulating the exhaust should be adjusted by the *quantity of coal* consumed. That is, the angular advance of their eccentric should, from time to time, be increased until at length the limit of greatest economy is attained. The result of course must not be judged by a computation of the indicator card, for that necessarily ignores the effect of the momentum of the reciprocating parts, since it measures the power in the *act* of being applied and not *subsequently* to the application, a distinction of great importance.

EQUALIZATION OF VALVE MOTION.

The cut-off being produced by a valve independent of that regulating the lead, its equalization may be accomplished without any of the difficulties incident to a single eccentric motion, where the slightest change in either event is immediately felt by the other. The desired result is approximately obtained for a single eccentric motion, by simply *lengthening* the cut-off valve stem an amount dependent on the ratio existing between crank and connecting rod.

If, however, the valve moves under the influence of a lever attached to the reciprocating parts of the engine, as in Fig. 52, the motion will become equalized by virtue of the irregularities thus introduced to the valve motion (a counterpart of those peculiar to the piston) and any slight inequalities of the main valve will be counteracted by a change in the length of the cut-off valve stem. The lead should be made *equal* for both faces of the main valve, and if neither the single eccentric nor link by which the latter is operated accomplishes an equalization of the exhaust closure, then a suitable amount of inside lap and clearance (as explained Part I) should be given to the exhaust edges of the valve face. The cut-off equality of the main valve, in itself considered, is of little consequence, hence the link should be arranged to give the *smallest* amount of slip attainable with an equality of the lead, and the cut-off should be regulated solely by the cut-off valves.

CLEARANCE.

This term is used to express, the extent of the space which exists between the piston at the extremity of its stroke and the valve face, or the cubic contents of the steam passage plus the unoccupied portion of the cylinder. Since for each stroke of the piston this space must be filled with steam, which in no way tends to improve the action of the parts, but rather increases the amount to be exhausted on the return stroke, it becomes desirable in long-stroke engines to locate the valve face as near the end of the stroke as possible, thus reducing the cubic contents of the passage and by its directness admitting the steam with a higher

initial pressure than could be obtained through a more tor-
tuous channel.

A convenient method for accomplishing this result, is to
separate the valve faces F and N by means of a stem, as
shown in Fig. 50, and forming them either in the shape of a

FIG. 50.

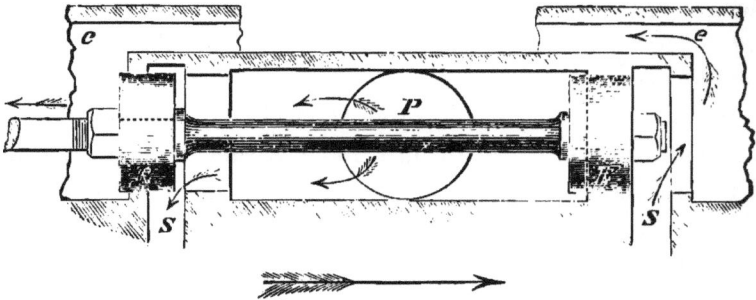

letter ⌐⌐ laid on its side, or as small pistons. The former
or D valve construction was much employed a few years
since, but has been gradually supplanted by the more mechanical arrangement of the piston valve. This, instead of
being surrounded by packing in the valve case, carries its
expansive packing in the same manner as an ordinary piston. The steam is received either between the pistons, from
a pipe P, giving the most direct admission to the cylinder,
or externally from chambers e e. The parts surrounding
the valves are well jacketed to prevent unequal expansion,
and slight irregularities are compensated by the elasticity
of the packing.

FRICTION.

The friction of a valve is quite independent of its surface,
except so far as the latter may increase the area upon
which the steam can exert an unbalanced pressure.

There are various expedients for relieving such pressure. Two of these have been illustrated in Figs. 48 and 50. A third consists in casting a standard in the exhaust port on which is bolted a concave plate, scraped to a bearing upon the inner surface of the valve, and holes drilled through the top admit the steam *under* the valve, which tends to relieve the external pressure. Besides these devices the valve is frequently mounted on steel rollers about 1¾ inches diameter resting on plates of the same metal, and thus the sliding converted into rolling friction. Heavy valves standing in a vertical position have additional rollers under their lower edge for supporting their great weight.

REDUCTION IN TRAVEL.

The work expended in friction depends directly upon the distance traversed by the valve. Hence, in marine engines every means possible should be employed to reduce this quantity to its minimum value. This object may be

FIG. 51.

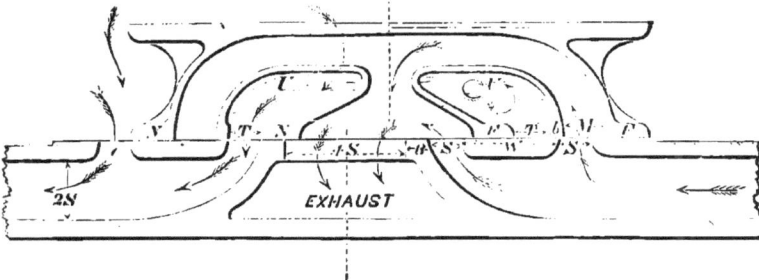

attained by increasing the number of the steam ports two or three fold, so that ½ or ⅓ of the travel will suffice to open the same extent of port area. For a reduction in the travel of one-half, the parts are commonly arranged as shown in Fig. 51. In this the valve edges F and N are separated by a distance wide enough to admit two steam passages U, U, whose openings T T communicate with the inner ports but not with the exhaust. They are formed with a width equal to that designed for the port opening. The general proportions of such valves may be concisely expressed as follows:

Total width of steam passage=2 S.
Each port=S.   Width of port opening=T.
Exhaust port=4 S.
Valve faces F and N=S+lap of valve.
Bridge W=½ travel+lap+T+¾ inch.
Width of exhaust bridge B dependent on thickness of cylinder.
Width of exhaust bridge b dependent on thickness of valve.

Occasionally the travel is reduced still further by inserting a third port beyond the other two. When so arranged the outer faces F and N are extended and through them passages cut, which admit the steam to the cylinder, but remain closed during the period of exhaust; consequently the *two* exhaust passages must be amply large to discharge the steam received through the *three* openings.

LINK AND RECIPROCATING MOTION COMBINED.

It was observed, while discussing the subject of link motion, that the central line of motion might be inclined at any angle to that of the piston motion, without affecting the character of the valve action. Fig. 52 illustrates this pecu-

liarity (for one position, namely, an angle of 75°) as applied
to a back-action engine having an independent cut-off.
The line E F represents the central line of the link motion.

Fig. 52.

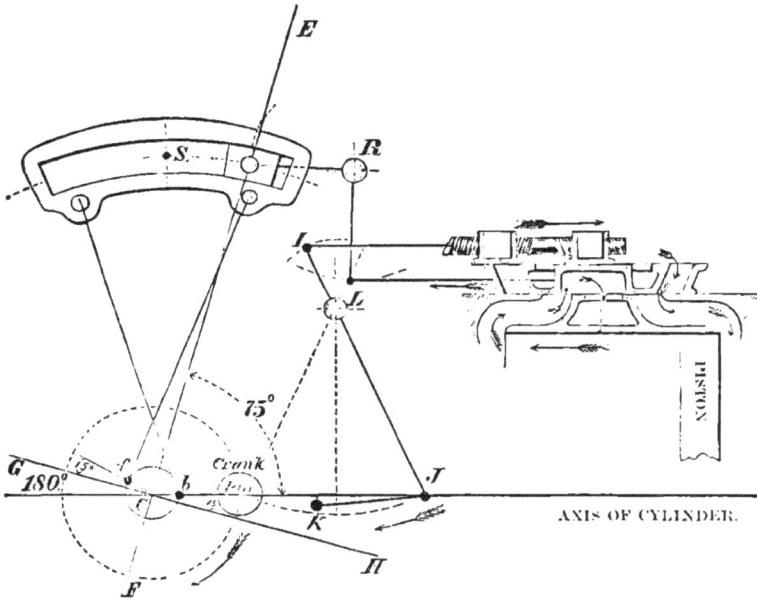

and G H a line at right angles thereto from which are laid
off the 15 degree angular advances of the eccentrics. The
eccentric rods being crossed, the lead of course diminishes
from the full to the mid gear of the link, but as this is only
employed in the full or immediately adjacent gears, the reduction proves advantageous, since it enables the engineer
to stop his engine by placing the link in the mid gear (see
Fig. 26). The cut-off valves are operated by a lever I J
connected through the link J K with the pin K, which is
secured to one of the piston rods or to the cross-head in a
direct-action engine.

This valve motion might have been derived from an eccentric with its normal position in the line E F, and acting on the valve through the medium of a rocker. The general conditions governing the motion have already been examined.

There are numerous other positions in which the link may be placed, and many other plans for connection with the valve, but it is believed that the typical forms have been presented with sufficient care and accuracy to enable the designer by their aid to accomplish any desired result.

It may prove a source of regret to some, that the numerous class of automatic cut-off gears, now so extensively applied to stationary engines, should have received no special attention in this Work. The Author however, has considered it most expedient to confine himself to general principles and to subjects requiring a solution in every day experience, leaving with those who hold such monopolies, and who *alone* can make use of them, the onus of explaining their principles and advertising their points of excellence.

Catalogue of the Scientific Publications

of D. Van Nostrand Company, 23 Murray

Street and 27 Warren Street, New York.

ADAMS, J. W. Sewers and Drains for Populous Dis-tricts. Embracing Rules and Formulas for the dimensions and construction of works of Sanitary Engineers. Fifth edition. 8vo, cloth..$2.50

ABBOTT, A. V. The Electrical Transmission of Ener-gy. A Manual for the Design of Electrical Circuits. Fully illustrated. 8vo, cloth...$4 50

ALEXANDER, J. H. Universal Dictionary of Weights and Measures, Ancient and Modern, reduced to the Standards of the United States of America. New edition, enlarged. 8vo, cloth, $3.50

—— **S. A. Broke Down: What Should I Do? A** Ready Reference and Key to Locomotive Engineers and Firemen, Round House Machinists, Conductors, Train Hands and Inspectors. With 5 folding plates. 12mo, cloth........................$1.50

ANDERSON, WILLIAM. On the Conversion of Heat into Work. A Practical Handbook on Heat-Engines. Third edition. Illustrated. 12mo, cloth................................$2.25

ARNOLD, Dr. Ammonia and Ammonium Compounds. A Practical Manual for Manufacturers, Chemists, Gas Engineers and Dyesetters. 12mo, cloth................................$2.00

ARNOLD, E. Armature Windings of Direct Current Dynamos. Extension and Application of a General Winding Rule. Translated from the original German by Francis B. DeGress, M. E. With numerous illustrations............................(In press.)

ATKINSON, PHILIP. The Elements of Electric Light-ing, including Electric Generation, Measurement, Storage, and Distribution. Eighth edition. Illustrated. 12mo, cloth.......$1.50

—— **The Elements of Dynamic Electricity and Mag-**netism. Second edition. 120 illustrations. 12mo, cloth....$2.00

—— **Elements of Static Electricity,** with full description of the Holtz and Topler Machines, and their mode of operating. Illustrated. 12mo, cloth....................................$1.50

—— **The Electric Transformation of Power and its Ap-**plication by the Electric Motor, including Electric Railway Construction. Illustrated. 12mo, cloth........................$2.00

AUCHINCLOSS, W. S. Link and Valve Motions
Simplified. Illustrated with 29 woodcuts and 20 lithographic plates,
together with a Travel Scale, and numerous useful tables. Thirteenth
edition. Revised. 8vo, cloth..................................$2.00

AXON, W. E. A. The Mechanic's Friend. A Collection
of Receipts and Practical Suggestions relating to Aquaria, Bronzing,
Cements, Drawing, Dyes, Electricity, Gilding, Glass-working, Glues,
Horology, Lacquers, Locomotives, Magnetism, Metal-working, Mod-
elling, Photography, Pyrotechny, Railways, Solders, Steam-Engine,
Telegraphy, Taxidermy, Varnishes, Waterproofing, and Miscellaneous
Tools, Instruments, Machines, and Processes connected with the
Chemical and Mechanic Arts. With numerous diagrams and woodcuts.
Fancy cloth...$1.50

BACON, F. W. A Treatise on the Richards Steam-
Engine Indicator, with directions for its use. By Charles T. Porter.
Revised, with notes and large additions as developed by American
practice ; with an appendix containing useful formulæ and rules for
engineers. Illustrated. Fourth edition. 12mo, cloth..........$1.00

BADT, F. B. Dynamo Tenders Hand-Book. With 140
illustrations. Second edition. 18mo, cloth..................$1.00

—— **Bell Hangers' Hand-Book. With 97 illustrations.**
18mo, cloth...$1.00

—— **Incandescent Wiring Hand-Book. With 35 illus-**
trations and five tables. 18mo, cloth......................$1.00

—— **Electric Transmission Hand-Book. With 22 illus-**
trations and 27 Tables. 18mo, cloth $1.00

BALCH, COL. GEO. T. Methods of Teaching Patriot-
ism in the Public Schools. 8vo, cloth......................$1.00

BALE, M. P. Pumps and Pumping. A Hand Book for
Pump Users. 12mo, cloth....................................$1.00

BARBA, J. The Use of Steel for Constructive Purposes.
Method of Working, Applying, and Testing Plates and Bars. With a
Preface by A. L. Holley, C.E. 12mo, cloth..................$1.50

BARNARD, F. A. P. Report on Machinery and Pro-
cesses of the Industrial Arts and Apparatus of the Exact Sciences at
the Paris Universal Exposition, 1867. 152 illustrations and 8 folding
plates. 8vo, cloth...$5.00

BEAUMONT, ROBERT. Color in Woven Design.
With 32 colored Plates and numerous original illustrations. Large,
12mo...$7.50

BECKWITH, ARTHUR. Pottery. Observations on
the Materials and Manufacture of Terra-Cotta, Stoneware, Fire-Brick,
Porcelain, Earthenware, Brick, Majolica, and Encaustic Tiles. 8vo,
paper. 2d edition....,..................... .... .60

BERNTHSEN, A. A Text-Book of Organic Chemistry.
Translated by George M'Gowan, Ph.D. Second English edition. Revised and extended by author and translator. Illustrated. 12mo, cloth $2.50

BIGGS, C. H. W. First Principles of Electrical
Engineering. Being an attempt to provide an Elementary Book for those intending to enter the profession of Electrical Engineering. Second edition. 12mo., cloth. Illustrated $1.00

BLAKE, W. P. Report upon the Precious Metals.
Being Statistical Notices of the principal Gold and Silver producing regions of the world, represented at the Paris Universal Exposition. 8vo, cloth.. $2.00

—— **Ceramic Art. A Report on Pottery, Porcelain,**
Tiles, Terra-Cotta, and Brick. 8vo, cloth.................... $2.00

BLAKESLEY, T. H. Alternating Currents of Elec-
tricity. For the use of Students and Engineers. Third edition, enlarged. 12mo, cloth.. $1.50

BLYTH, A. WYNTER, M.R.C.S., F.C.S. Foods: their
Composition and Analysis. Third edition Crown 8vo, cloth. ..$6.00

—— **Poisons: their Effects and Detection.** Third edition.
Revised and enlarged. Crown 8vo, cloth..................... $7.50

BODMER, G. R. Hydraulic Motors; Turbines and
Pressure Engines, for the use of Engineers, Manufacturers, and Students. Second edition, revised and enlarged. With 204 illustrations. 12mo, cloth................. .... $5.00

BOTTONE, S. R. Electrical Instrument Making for
Amateurs. A Practical Handbook. With 48 illustrations. 12mo, cloth. Fifth Edition. Revised........................ .... .50

—— **Electric Bells, and all about them. A Practical**
Book for Practical Men. With more than 100 illustrations. 12mo, cloth. Fourth Edition. Revised and Enlarged............... .50

—— **The Dynamo: How Made and How Used. A**
Book for Amateurs. Eighth edition. 12mo, cloth........... $1.00

—— **Electro Motors: How Made and How Used. A**
Hand Book for Amateurs and Practical Men. Second edition. 12mo, cloth .... .50

BONNEY, G. E. The Electro-Platers' Hand Book.
A Manual for Amateurs and Young Students on Electro-Metallurgy. 60 Illustrations. 12mo, cloth............................ $1.20

BOW, R. H. A Treatise on Bracing. With its applica-
tion to Bridges and other Structures of Wood or Iron. 156 illustrations. 8vo, cloth................................... $1.50

BOWSER, Prof. E. A. An Elementary Treatise on
Analytic Geometry. Embracing Plane Geometry, and an Introduction to Geometry of three Dimensions. 12mo, cloth. Eighteenth
edition $1.75

—— **An Elementary Treatise on the Differential and**
Integral Calculus. With numerous examples. 12mo, cloth. Four-
teenth edition $2.25

—— **An Elementary Treatise on Analytic Mechanics.**
With numerous examples. 12mo, cloth. Eighth edition. $3.00

BOWSER, Prof. E. A. Academic Algebra Second edi-
tion, 12mo, cloth. $1.25

—— **An Elementary Treatise on Hydro-Mechanics.**
With numerous examples. 12mo, cloth. Fourth edition. $2.50

—— **College Algebra.** Fourth edition. 12mo, cloth $1.75

—— **Elements of Plane and Solid Geometry. 12mo,**
cloth. Second edition. $1.40

BOWIE, AUG. J., Jun., M. E. A Practical Treatise on
Hydraulic Mining in California. With Description of the Use and
Construction of Ditches, Flumes, Wrought-iron Pipes and Dams ;
Flow of Water on Heavy Grades, and its Applicability, under High
Pressure, to Mining. Fifth edition. Small quarto, cloth. Illus-
trated. $5.00

BURGH, N. P. Modern Marine Engineering, applied
to Paddle and Screw Propulsion. Consisting of 36 colored plates, 259
practical woodcut illustrations, and 403 pages of descriptive matter.
The whole being an exposition of the present practice of James Watt
& Co., J. & G. Rennie, R. Napier & Sons, and other celebrated firms.
Thick 4to vol., half morocco. $10.00

BURT, W. A. Key to the Solar Compass, and Survey-
or's Companion. Comprising all the rules necessary for use in the
field ; also description of the Linear Surveys and Public Land System
of the United States, Notes on the Barometer, Suggestions for an Out-
fit for a Survey of Four Months, etc. Fifth edition. Pocket-book
form, tuck. $2.50

CALDWELL, Prof. GEO. C., and BRENEMAN, Prof.
A. A. Manual of Introductory Chemical Practice. For the use of
Students in Colleges and Normal and High Schools. Fourth edition,
revised and corrected. 8vo, cloth. Illustrated. $1.50

CAMPIN, FRANCIS. On the Construction of Iron
Roofs. A Theoretical and Practical Treatise, with wood-cuts and
Plates of Roofs recently executed. 8vo, cloth. $2.00

CARTER, E. T. Motive Power and Gearing for Elec-
trical Machinery . . . .(In press.)

CHAUVENET, Prof. W. New Method of Correcting Lunar Distances, and Improved Method of Finding the Error and Rate of a Chronometer, by Equal Altitudes. 8vo, cloth........$2 00

CHURCH, JOHN A. Notes of a Metallurgical Journey in Europe. 8vo, cloth.................................$2.00

CLARK, D. KINNEAR, C.E. A Manual of Rules, Tables and Data for Mechanical Engineers. Based on the most recent investigations. Illustrated with numerous diagrams. 1,012 pages. 8vo, cloth, Sixth edition..................................$5.00
Half morocco..$7.50

—— **Fuel; its Combustion and Economy, consisting of** abridgements of Treatise on the Combustion of Coal. By C. W. Williams; and the Economy of Fuel, by T. S. Prideaux. With extensive additions in recent practice in the Combustion and Economy of Fuel, Coal, Coke, Wood, Peat, Petroleum, etc. Fourth edition. 12mo, cloth..... ....$1.50

—— **The Mechanical Engineer's Pocket Book of** Tables, Formulæ, Rules and Data. A Handy Book of Reference for Daily Use in Engineering Practice 16mo., morocco. Second edition...................................,..........$3.00

—— **Tramways, their Construction and Working, em-**bracing a comprehensive history of the system, with accounts of the various modes of traction, a description of the varieties of rolling stock, and ample details of Cost and Working Expenses. *Second edition.* Re-written and greatly enlarged, with upwards of 400 illustrations. Thick 8vo, cloth.................................$9 00

CLARK, JACOB M. A new System of Laying Out Railway Turn-outs instantly, by inspection from tables. 12mo, leatherette..$1.00

CLEEMANN, THOS. M. The Railroad Engineer's Practice. Being a Short but Complete Description of the Duties of the Young Engineer in the Prelimary and Location Surveys and in Construction. 4th ed., revised and enlarged. Illus. 12mo, cloth $1.50

CLEVENGER, S. R. A Treatise on the Method of Government Surveying as prescribed by the U. S. Congress and Commissioner of the General Land Office, with complete Mathematical, Astronomical and Practical Instructions for the use of the United States Surveyors in the field. 16mo, morocco.........$2.50

COLBURN, ZERAH. The Gas-Works of London. 12mo, boards... .60

COLLINS, JAS. E. The Private Book of Useful Alloys and Memoranda for Goldsmiths, Jewelers, etc. 18mo, cloth..... .50

CORNWALL, Prof. H. B. Manual of Blow-Pipe An-alysis, Qualitative and Quantitative. With a Complete System of Descriptive Mineralogy. 8vo, cloth, with many illustrations....$2.50

CRAIG, B. F. Weights and Measures. An account of the Decimal System, with Tables of Conversion for Commercial and Scientific Uses. Square 32mo, limp cloth........ .... .50

CROCKER, F. B. Electric Lighting. Designed as a Text-Book for Colleges and Technical Schools, and as a handbook for Engineers, Architects and others interested in the Installation or Operation of Electric Lighting Plants........................(In press)

CROCKER, F. B., and S. S. WHEELER. The Practical Management of Dynamos and Motors. Third edition, (fifth thousand) revised and enlarged, with a special chapter by H. A. Foster. 12mo, cloth, illustrated................................. ....$1 00

CUMMING, LINNÆUS, M.A. Electricity treated Experimentally. For the use of Schools and Students. New edition. 12mo, cloth...$1.50

DAVIES, E. H. Machinery for Metalliferous Mines. A Practical Treatise for Mining Engineers, Metallurgists and Manufacturers. With upwards of 300 illustrations. 8vo, cloth......$5.00

DAVIS, JOHN W., C.E. Formulæ for the Calculation of Rail Road Excavation and Embankment, and for finding Average Haul. Second edition. Octavo, half roan....................$1.50

DAVIS, J. WOODBRIDGE. Theoretical Astronomy. Dynamics of the Sun. 4to, paper...........................$3.00

DIXON, D. B. The Machinist's and Steam Engineer's Practical Calculator. A Compilation of Useful Rules and Problems arithmetically solved, together with General Information applicable to Shop-Tools, Mill-Gearing, Pulleys and Shafts, Steam-Boilers and Engines. Embracing valuable Tables and Instruction in Screwcutting, Valve and Link Motion, etc. 16mo, full morocco, pocket form..$2.00

DODD, GEO. Dictionary of Manufactures, Mining, Machinery, and the Industrial Arts. 12mo, cloth.............$1.50

DORR, B. F. The Surveyor's Guide and Pocket Table Book. 18mo, morocco flaps. Third Edition.................$2.00

DUBOIS, A. J. The New Method of Graphic Statics. With 60 illustrations. 8vo, cloth...........................$1.50

EDDY, Prof. H. T. Researches in Graphical Statics. Embracing New Constructions in Graphical Statics, a New General Method in Graphical Statics, and the Theory of Internal Stress in Graphical Statics. 8vo, cloth..............................$1.50

—— **Maximum Stresses under Concentrated Loads.** Treated graphically. Illustrated. 8vo, cloth $1.50

EISSLER, M. The Metallurgy of Gold; a Practical Treatise on the Metallurgical Treatment of Gold-Bearing Ores, including the Processes of Concentration and Chlorination, and the Assaying, Melting and Refining of Gold. Fourth edition, revised and greatly enlarged. 187 illustrations. 12mo, cl.................$7.50

EISSLER, M. **The Metallurgy of Silver; a practical** treatise on the Amalgamation, Roasting and Lixiviation of Silver Ores, including the Assaying, Melting and Refining of Silver Bullion. 124 illustrations. Second edition, enlarged. 12mo, cloth.........$4.00

—— **The Metallurgy of Argentiferous Lead; a Prac-** tical Treatise on the Smelting of Silver-Lead Ores and the Refining of Lead Bullion. Including Reports on Various Smelting Establishments and Descriptions of Modern Smelting Furnaces and Plants in Europe and America. With 183 illustrations. 8vo, cloth......$5.00

ELIOT, Prof. C. W., and STORER, Prof. F. H. **A** Compendious Manual of Qualitative Chemical Analysis. Revised with the co-operation of the authors, by Prof. William R. Nichols. Illustrated. Seventeenth edition, newly revised by Prof. W. B. Lindsay. 12mo, cloth. $1.50

ELLIOT, Maj. GEO. H., U. S. E. **European Light-** House Systems. Being a Report of a Tour of Inspection made in 1873. 51 engravings and 21 woodcuts. 8vo, cloth...........$5.00

ENDLICH, F. M. **Manual of Qualitative Blow-Pipe** Analysis and Determinative Mineralogy. Illustrations and Colored Plate of Spectra. 8vo, cloth................................$4.00

EVERETT, J. D. **Elementary Text-Book of Physics.** Illustrated. Seventh edition. 12mo, cloth..................$1.40

EWING, Prof. A. J. **The Magnetic Induction in Iron** and other metals. 159 illustrations. 8vo, cloth $4 00

FANNING, J. T. **A Practical Treatise on Hydraulic** and Water-Supply Engineering. Relating to the Hydrology, Hydrodynamics, and Practical Construction of Water-Works in Nort America. 180 illustrations. 8vo, cloth. Eleventh edition, revised, enlarged, and new tables and illustrations added. 650 pages....$5.00

FISH, J. C. L. **Lettering of Working Drawings.** Thirteen plates, with descriptive text. Oblong, 9x12½, boards......$1.00

FISKE, Lieut. BRADLEY A., U.S.N. **Electricity in** Theory and Practice; or, The Elements of Electrical Engineering. Eighth edition. 8vo, cloth. $2.50

FLEMING, Prof. J. A. **The Alternate Current Trans-** former in Theory and Practice. Vol. I—The Induction of Electric Currents; 500 pages. Second edition. Illustrated. 8vo, cloth.$3.00 Vol. II. The Utilization of Induced Currents. Illustrated. 8vo, cloth..$5.00

FLEMING, Prof. J. A. **Electric Lamps and Electric** Lighting. Being a course of four lectures delivered at the Royal Institution, April-May, 1894. 8vo, cloth, fully illustrated........$3.00

FLEMING, Prof. J. A. **Electrical Laboratory Notes and** Forms, elementary and advanced. 4to, cloth, illustrated......$5.00

FOLEY, (NELSON) and THOS. PRAY, Jr. The
Mechanical Engineers' Reference Book for Machine and Boiler Construction, in two parts. Part 1—General Engineering Data. Part 2
—Boiler Construction. With fifty-one plates and numerous illustrations, specially drawn for this work. Folio, half mor.........$25.00

FORNEY, MATTHIAS N. Catechism of the Locomo-
tive. Second edition, revised and enlarged. Forty-sixth thousand.
8vo, cloth $3.50

FOSTER, Gen. J. G., U.S.A. Submarine Blasting in
Boston Harbor, Massachusetts. Removal of Tower and Corwin
Rocks. Illustrated with 7 plates. 4to, cloth.................$3.50

FRANCIS, Jas. B., C.E. Lowell Hydraulic Experiments. Being a selection from experiments on Hydraulic Motors,
on the Flow of Water over Weirs, in open Canals of uniform rec-
tangular section, and through submerged Orifices and diverging
Tubes. Made at Lowell, Mass. Fourth edition, revised and
enlarged, with many new experiments, and illustrated with 23 copper-
plate engravings. 4to, cloth.............................$15.00

GEIPEL, WM. and KILGOUR, M. H. A Pocketbook
of Electrical Engineering Formulæ. Illustrated. 18mo, mor..$3.00
Large paper edition, wide margins. 8vo, morocco, gilt edges $5.00

GERBER, NICHOLAS. Chemical and Physical Analysis of Milk, Condensed Milk and Infant's Milk-Food. 8vo,
cloth..$1.25

GILLMORE, Gen. Q. A. Treatise on Limes, Hydraulic
Cements, and Mortars. Papers on Practical Engineering, United
States Engineer Department, No. 9, containing Reports of numerous
Experiments conducted in New York City during the years 1858 to
1861, inclusive. With numerous illustrations. 8vo, cloth......$4.00

—— Practical Treatise on the Construction of Roads,
Streets, and Pavements. With 70 illustrations. 12mo, cloth, $2.00

—— Report on Strength of the Building-Stones in the
United States, etc. 8vo, illustrated, cloth...................$1.00

GOODEVE, T. M. A Text-Book on the Steam-Engine.
With a Supplement on Gas-Engines. Twelfth edition. Enlarged.
143 illustrations. 12mo, cloth............................$2.00

GORDON, J. E. H. School Electricity. Illustrations.
12mo, cloth...$2.00

GORE, G., F. R. S. The Art of Electrolytic Separation of Metals, etc. (Theoretical and Practical.) Illustrated. 8vo,
cloth..$3.50

—— Electro Chemistry Inorganic. 2d edition. 8vo, cloth.
....$ 80

GRAY, JOHN, B.Sc. Electrical Influence Machines.
A full account of their historical development and modern forms, with instructions for making them. 89 illustrations and 3 folding plates. 12mo, cloth..$1.75

GRIFFITHS, A. B., Ph.D. A Treatise on Manures,
or the Philosophy of Manuring. A Practical Hand-Book for the Agriculturist, Manufacturer, and Student. 12mo, cloth........$3.00

GRUNER, M. L. The Manufacture of Steel. Translated from the French, by Lenox Smith; with an appendix on the Bessemer process in the United States, by the translator. Illustrated. 8vo, cloth $3.50

GURDEN, RICHARD LLOYD. Traverse Tables:
computed to 4 places Decimals for every ° of angle up to 100 of Distance. For the use of Surveyors and Engineers. New edition. Folio, half mo..$7.50

GUY, ARTHUR F. Electric Light and Power, giving
the Result of Practical Experience in Central-Station Work. 8vo, cloth. Illustrated $2 50

HAEDER, HERMAN, C. E. A Handbook on the
Steam Engine. With especial reference to small and medium sized engines. English edition, re-edited by the author from the second German edition, and translated with considerable additions and alterations by H. H. P. Powles. 12mo, cloth. Nearly 1100 illus. ..$3.00

HALSEY, F. A. Slide Valve Gears, an Explanation
of the action and Construction of Plain and Cut-off Slide Valves. Illustrated. 12mo, cloth. Third edition $1.50

HAMILTON, W. G. Useful Information for Railway
Men. Tenth edition, revised and enlarged. 562 pages, pocket form. Morocco, gilt..$2.00

HANCOCK, HERBERT. Text-Book of Mechanics and
Hydrostatics, with over 500 diagrams. 8vo, cloth $1 75

HARRISON, W. B. The Mechanics' Tool Book.
With Practical Rules and Suggestions for use of Machinists, Iron-Workers, and others. Illustrated with 44 engravings. 12mo, cloth,..$1.50

HASKINS, C. H. The Galvanometer and its Uses.
A Manual for Electricians and Students. Fourth edition. 12mo, cloth..$1.50

HEAP, Major D. P., U. S. A. Electrical Appliances of
the Present Day. Report of the Paris Electrical Exposition of 1881. 250 illustrations. 8vo, cloth............................$2.00

HEAVSIDE, OLIVER. Electromagnetic Theory. Vol. 1. 8vo, cloth..$5 00

HENRICI, OLAUS. Skeleton Structures, Applied to the Building of Steel and Iron Bridges. Illustrated.........$1.50

HERRMANN, Gustav. The Graphical Statics of Mechanism. A Guide for the Use of Machinists, Architects, and Engineers ; and also a Text-book for Technical Schools. Translated and annotated by A. P. Smith, M. E. 12mo, cloth, 7 folding plates. Second edition....................................$2.00

HEWSON, WM. Principles and Practice of embanking Lands from River Floods, as applied to the Levees of the Mississippi. 8vo, cloth..$2.00

HOBBS, W. R. P. The Arithmetic of Electrical Measurements, with numerous examples. Fully Worked. 12mo, cloth, .50

HOFFMAN, H. D. The Metallurgy of Lead and the Desilverization of Base Bullion. 275 Illustrations. 8vo, cloth. $6.00

HOLLEY, ALEXANDER L. Railway Practice. American and European Railway practice in the Economical Generation of Steam, including the Materials and Construction of Coal-burning Boilers, Combustion, the Variable Blast, Vaporization, Circulation, Superheating, Supplying and Heating Feed Water, etc., and the Adaptation of Wood and Coke-burning Engines to Coal-burning ; and in Permanent Way, including Road-bed, Sleepers, Rails, Joint Fastenings, Street Railways, etc. With 77 lithographed plates. Folio, cloth.. $12.00

HOLMES, A. BROMLEY. The Electric Light Popularly Explained. Fifth edition. Illustrated. 12mo, paper.... .40

HOSPITALIER, E. Polyphased Alternating Currents. Illustrated. 8vo, cloth...................................$1.40

HOWARD, C. R. Earthwork Mensuration on the Basis of the Prismoidal Formulæ. Containing Simple and Labor-saving Method of obtaining Prismoidal Contents directly from End Areas. Illustrated by Examples and accompanied by Plain Rules for Practical Uses. Illustrated. 8vo, cloth...........................$1.50

HOWE, HENRY M. The Metallurgy of Steel. Vol. I. Third edition, revised and enlarged. 4to, cloth...........$10.00

HUMBER, WILLIAM, C. E. A Handy Book for the Calculation of Strains in Girders, and Similar Structures, and their Strength ; Consisting of Formulæ and Corresponding Diagrams, with numerous details for practical application, etc. Fourth edition. 12mo, cloth... ....$2.50

HUTTON, W. S. Steam Boiler Construction. A Practical Hand Book for Engineers, Boiler Makers and Steam Users. With upwards of 300 illustrations. Second edition. 8vo, cloth, $7.00

HUTTON, W. S. Practical Engineer's Hand-Book, comprising a treatise on Modern Engines and Boilers, Marine, Locomotive and Stationary. Fourth edition. Carefully revised, with additions. With upwards of 570 illustrations. 8vo, cloth, ...$7 00

—— **The Works' Manager's Hand-Book of Modern** Rules, Tables, and Data for Civil and Mechanical Engineers, Millwrights and Boiler-makers, etc., etc. With upwards of 150 illustrations. Fifth edition. Carefully revised, with additions. 8vo, cloth, ...$6 00

ISHERWOOD, B. F. Engineering Precedents for Steam Machinery. Arranged in the most practical and useful manner for Engineers. With illustrations. 2 vols. in 1. 8vo, cloth.....$2.50

JAMIESON, ANDREW, C.E. A Text-Book on Steam and Steam-Engines. Specially arranged for the use of Science and Art, City and Guilds of London Institute, and other Engineering Students. Tenth edition. Illustrated. 12mo, cloth...........$3.00

—— **Elementary Manual on Steam and the Steam-** Engine. Specially arranged for the use of First-Year Science and Art, City and Guilds of London Institute, and other Elementary Engineering Students. Third edition. 12mo, cloth $1.40

JANNETTAZ, EDWARD. A Guide to the Determina- tion of Rocks : being an Introduction to Lithology. Translated from the French by G. W. Plympton, Professor of Physical Science at Brooklyn Polytechnic Institute. 12mo, cloth....... ....$1.50

JOYNSON, F. H. The Metals used in Construction. Iron, Steel, Bessemer Metal, etc. Illustrated. 12mo, cloth.... .75

—— **Designing and Construction of Machine Gearing.** Illustrated. 8vo, cloth $2.00

KANSAS CITY BIDGE, THE With an Account of the Regimen of the Missouri River and a Description of the Methods used for Founding in that River. By O. Channte, Chief Engineer, and George Morrison, Assistant Engineer. Illustrated with 5 lithographic views and 12 plates of plans. 4to, cloth....................$6.00

KAPP, GISBERT, C.E. Electric Transmission of Ener- gy and its Transformation, Subdivision, and Distribution. A Practical hand-book. Fourth edition. Revised. 12mo, cloth.. ....$3.50

—— **Dynamos, Alternators and Transformers.** 138 Illustrations. 12mo, cloth...$4.00

KEMPE, H. R. The Electrical Engineer's Pocket Book of Modern Rules, Formulæ, Tables and Data. Illustrated. 32mo. Mor. gilt....................................$1.75

KENNELLY, A. E. Theoretical Elements of Electro- Dynamic Machinery. 8vo, cloth $1.50

KILGOUR, M. H., SWAN, H., and BIGGS, C. H. W.
Electrical Distribution, its Theory and Practice. 174 Illustrations.
12mo, cloth..$4.00

KING, W. H.   Lessons and Practical Notes on Steam.
The Steam-Engine, Propellers, etc., for Young Marine Engineers,
Students, and others.   Revised by Chief Engineer J. W. King, United
States Navy.   Nineteenth edition, enlarged.   8vo, cloth.........$2.00

KIRKALDY, Wm. G.   Illustrations of David Kirk-
aldy's System of Mechanical Testing, as Originated and Carried On
by him during a Quarter of a Century.   Comprising a Large Selection
of Tabulated Results, showing the Strength and other Properties of
Materials used in Construction, with Explanatory Text and Historical
Sketch.   Numerous engravings and 25 lithographed plates.   4to,
cloth.. ....$20.00

KIRKWOOD, JAS. P.   Report on the Filtration of
River Waters for the supply of Cities, as practised in Europe, made
to the Board of Water Commissioners of the city of St. Louis.   Illustrated by 30 double-plate engravings.   4to, cloth..............$15.00

KUNZ, GEO. F.   Gems and Precious Stones of North
America.   A Popular Description of Their Occurence, Value, History,
Archæology, and of the Collections in which They Exist; also a Chap-
ter on Pearls and on Remarkable Foreign Gems owned in the United
States.   Illustrated with eight Colored Plates and numerous minor
Engravings.   Second edition, with appendix.   4to, cloth......$10.00

LARRABEE, C. S.   Cipher and Secret Letter and Tele-
graphic Code, with Hog's Improvements.   The most perfect Secret
Code ever invented or discovered.   Impossible to read without the
key.   18mo, cloth 60

LAZELLE, H. M.   One Law in Nature.   A New
Corpuscular Theory comprehending Unity of Force, Identity of
Matter, and its Multiple Atom Constitution, etc.   12mo, cloth,..$1.50

LEITZE, ERNST.   Modern Heliographic Processes.
A Manual of Instruction in the Art of Reproducing Drawings, Engravings, etc., by the action of Light.   With 32 woodcuts and ten
specimens of Heliograms.   8vo, cloth.   Second edition........$3.00

LOCKE, ALFRED G. and CHARLES G.   A Practical
Treatise on the Manufacture of Sulphuric Acid.   With 77 Construc-
tive Plates drawn to Scale Measurements, and other Illustrations.
Royal 8vo, cloth...$15.00

LOCKWOOD, THOS. D.   Electricity, Magnetism, and
Electro-Telegraphy.   A Practical Guide for Students, Operators, and
Inspectors.   8vo, cloth.   Third edition......................$2.50

LOCKWOOD, THOS. D. Electrical Measurement and the Galvometer ; its Construction and Uses. Second edition. 32 illustrations. 12mo, cloth.......................................$1.50

LODGE, OLIVER J. Elementary Mechanics, includ- ing Hydrostatics and Pneumatics. Revised edition. 12mo, cloth, $1.20

LORING, A. E. A Hand-Book of the Electro-Magnetic Telegraph. Paper boards.................................. .50
Cloth.. .75
Morocco...$1.00

LOVELL, D. H. Practical Switch Work. A Hand- Book for Track Foremen. Illustrated. 12mo, cloth...........$1.50

LUNGE, GEO. A Theoretical and Practical Treatise on the Manufacture of Sulphuric Acid and Alkali with the Collateral Branches. Vol. I. Sulphuric Acid. Second edition, revised and enlarged. 342 illustrations. 8vo, cloth.. ....$15.00
Vol. II. 8vo, cloth......................................$16.80
Vol. III. 8vo, cloth.................................... 9.60

—— and HUNTER, F. The Alkali Maker's Pocket Book. Tables and Analytical Methods for Manufacturers of Sulphuric Acid, Nitric Acid, Soda, Potash and Ammonia. Second edition. 12mo, cloth......................................$3.00

MACCORD, Prof. C. W. A Practical Treatise on the Slide-Valve by Eccentrics, examining by methods the action of the Eccentric upon the Slide-Valve, and explaining the practical processes of laying out the movements, adapting the Valve for its various duties in the Steam-Engine. Second edition. Illustrated. 4to, cloth $2.50

MAYER, PROF. A. M. Lecture Notes on Physics. 8vo, cloth..$2.00

McCULLOCH, Prof. R. S. Elementary Treatise on the Mechanical Theory of Heat, and its application to Air and Steam Engines. 8vo, cloth...................................$3.50

MERRILL, Col. WM. E., U.S.A. Iron Truss Bridges for Railroads. The method of calculating strains in Trusses, with a careful comparison of the most prominent Trusses, in reference to economy in combination, etc. Illustrated. 4to, cloth. 4th ed., $5.00

METAL TURNING. By a Foreman Pattern Maker. Illustrated with 81 engravings. 12mo, cloth.................$1.50

MINIFIE, WM. Mechanical Drawing. A Textbook of Geometrical Drawing for the use of Mechanics and Schools, in which the Definitions and Rules of Geometry are familiarly explained ; the Practical Problems are arranged from the most simple to the more complex, and in their description technicalities are avoided as much as possible. With illustrations for Drawing Plans, Sections, and Elevations of Railways and Machinery ; an Introduction to Isometrical Drawing, and an Essay on Linear Perspective and Shadows. Illustrated with over 200 diagrams engraved on steel. Ninth thousand. With an appendix on the Theory and Application of Colors. 8vo, cloth..$4.00

MINIFIE, WM. Geometrical Drawing. Abridged from the octavo edition, for the use of schools. Illustrated with 48 steel plates. Ninth edition. 12mo, cloth.........................$2.00

MODERN METEOROLOGY. A Series of Six Lectures, delivered under the auspices of the Meteorological Society in 1870. Illustrated. 12mo, cloth.................................$1.58

MOONEY, WM. The American Gas Engineers' and Superintendents' Hand Book, consisting of Rules, Reference Tables and original matter pertaining to the Manufacture, Manipulation and Distribution of Illuminating Gas. Illustrated. 12mo, morocco.$3.00

MORRIS, E. Easy Rules for the Measurement of Earth- works by means of the Prismoidal Formula. 8vo, cloth, illns'td..$1.50

MOSES, ALFRED J., and PARSONS, C. L. Elements of Mineralogy, Crystallography and Blowpipe Analysis from a practical standpoint. 8vo, cloth, 336 illustrations................$2.00

MOTT, H. A., Jun. A Practical Treatise on Chemistry (Qualitative and Quantitative Analysis), Stoichiometry, Blow-pipe Analysis, Mineralogy, Assaying, Pharmaceutical Preparations, Human Secretions, Specific Gravities, Weights and Measures, etc. Second revised edition, 650 pages. 8vo, cloth....,.................$4.00

MULLIN, JOSEPH P., M.E. Modern Moulding and Pattern-Making. A Practical Treatise upon Pattern-Shop and Foundry Work : embracing the Moulding of Pulleys, Spur Gears, Worm Gears, Balance-Wheels, Stationary Engine and Locomotive Cylinders, Globe Valves, Tool Work, Mining Machinery, Screw Propellers, Pattern-Shop Machinery, and the latest improvements in English and American Cupolas ; together with a large collection of original and carefully selected Rules and Tables for every-day use in the Drawing Office, Pattern-Shop and Foundry. 12mo, cloth, illustrated....$2.50

MUNRO, JOHN, C.E., and JAMIESON, ANDREW, C.E. A Pocketbook of Electrical Rules and Tables for the 'use of Electricians and Engineers. Tenth edition, revised and enlarged. With numerous diagrams. Pocket size. Leather............$2.50

MURPHY, J. G., M.E. Practical Mining. A Field Manual for Mining Engineers. With Hints for Investors in Mining Properties. 16mo, morocco tucks.......................$1.50

NAQUET, A. Legal Chemistry. A Guide to the Detec- tion of Poisons, Falsification of Writings, Adulteration of Alimentary and Pharmaceutical Substances, Analysis of Ashes, and examination of Hair, Coins, Arms, and Stains, as applied to Chemical Jurisprudence, for the use of Chemists, Physicians, Lawyers, Pharmacists and Experts. Translated, with additions, including a list of books and memoirs on Toxicology, etc., from the French, by J. P. Battershall, Ph.D., with a preface by C. F. Chandler, Ph.D., M.D., LL.D. 12mo, cloth..$2.00

NEWALL, JOHN W. Plain Practical Directions for Drawing, Sizing and Cutting Bevel-Gears, showing how the Teeth may be cut in a Plain Milling Machine or Gear Cutter so as to give them a correct shape from end to end; and showing how to get out all particulars for the Workshop without making any Drawings. Including a Full Set of Tables of Reference. Folding Plates. 8vo, cloth..$3.00

NEWLANDS, JAMES. The Carpenters' and Joiners' Assistant : being a Comprehensive Treatise on the Selection, Preparation and Strength of Materials, and the Mechanical Principles of Framing, with their application in Carpentry, Joinery, and Hand-Railing ; also, a Complete Treatise on Sines ; and an illustrated Glossary of Terms used in Architecture and Building. Illustrated. Folio, half mor...$15.00

NIPHER, FRANCIS E., A.M. Theory of Magnetic Measurements, with an appendix on the Method of Least Squares. 12mo, cloth..$1.00

NOAD, HENRY M. The Students' Text Book of Electricity. A new edition, carefully revised. With an Introduction and additional chapters by W. H. Preece. With 471 illustrations. 12mo, cloth $4.00

NUGENT, E. Treatise on Optics; or, Light and Sight theoretically and practically treated, with the application to Fine Art and Industrial Pursuits. With 103 illustrations. 12mo, cloth...$1.50

PAGE, DAVID. The Earth's Crust, A Handy Outline of Geology. 16mo, cloth...75

PALAZ, A., ScD. A Treatise on Industrial Photometry, with special application to Electric Lighting. Authorized translation from the French by George W. Patterson, Jr. 8vo, cloth. Illustrated..$4.00

PARSHALL, H. F., and HOBART, H. M. Armature Windings of Electric Machines. With 140 full page plates, 65 tables, and 165 pages of descriptive letter-press. 4to, cloth......$7 50

PEIRCE, B. System of Analytic Mechanics. 4to, cloth...$10.00

—— Linear Associative Algebra. New edition with addenda and notes by C. L. Pierce. 4to, cloth $5.00

PETERS, Dr. EDWARD M. Modern American Methods of Copper Smelting. Numerous illustrations. Sixth edition, revised and enlarged. 8vo, cloth............................$4.00

PHILLIPS, JOSHUA. Engineering Chemistry. A Practical Treatise for the use of Analytical Chemists, Engineers, Iron Masters, Iron Founders, students and others. Comprising methods of Analysis and Valuation of the principal materials used in Engineering works, with numerous Analysis, Examples and Suggestions. 314 Illus. Second edition, revised and enlarged. 8vo, cloth...$4.00

PLANE TABLE, The. Its Uses in Topographical
Surveying. From the Papers of the United States Coast Survey.
Illustrated. 8vo, cloth..................................$2.00
"This work gives a description of the Plane Table employed at the
United States Coast Survey office, and the manner of using it."

PLANTE, GASTON. The Storage of Electrical Energy,
and Researches in the Effects created by Currents, combining Quantity with High Tension. Translated from the French by Paul B.
Elwell. 89 illustrations. 8vo.... ....$4.00

PLATTNER. Manual of Qualitative and Quantitative
Analysis with the Blow-Pipe. From the last German edition, revised
and enlarged, by Prof. Th. Richter, of the Royal Saxon Mining
Academy. Translated by Prof. H. B. Cornwall, assisted by John H.
Caswell. Illustrated with 87 woodcuts and one lithographic plate.
Seventh edition, revised. 560 pages. 8vo, cloth............. $5.00

PLYMPTON, Prof GEO. W. The Blow-Pipe. A
Guide to its use in the Determination of Salts and Minerals. Compiled from various sources. 12mo, cloth.....................$1.50

—— **The Aneroid Barometer: its Construction and Use.**
Compiled from several sources. Fourth edition. 16mo, boards, il-
lustrated... .50
Morocco, ...$1.00

POCKET LOGARITHMS, to Four Places of Decimals,
including Logarithms of Numbers, and Logarithmic Sines and Tangents to Single Minutes. To which is added a Table of Natural
Sines, Tangents, and Co-Tangents. 16mo, boards............. .50

POOLE, JOSEPH. The Practical Telephone Hand-
Book and Guide to the Telephonic Exchange. 227 illustrations.
12mo, cloth..:.....$1.00

POPE, F. L. Modern Practice of the Electric Tele-
graph. A Technical Hand-Book for Electricians, Managers and
Operators. Fourteenth edition, rewritten and enlarged, and fully
illustrated. 8vo, cloth......................................$1.50

PRAY, Jr., THOMAS. Twenty Years with the In-
dicator; being a Practical Text-Book for the Engineer or the Student,
with no complex Formulæ. Illustrated. 8vo, cloth...........$2.50

—— **Steam Tables and Engine Constants. Compiled**
from Regnault, Rankine and Dixon directly, making use of the exact
records. 8vo, cloth $2 00

PRACTICAL IRON FOUNDING. By the author of
"Pattern Making," &c., &c. Illustrated with over one hundred
engravings. 12mo, Cloth.....................................$1.50

PREECE, W. H., and STUBBS, A. T. Manual of Tele-
phony. Illustrations and plates. 12mo, cloth...............$4.50

PRESCOTT, Prof. A. B. Organic Analysis. A Manual of the Descriptive and Analytical Chemistry of certain Carbon Compounds in Common Use ; a Guide in the Qualitative and Quantitative Analysis of Organic Materials in Commercial and Pharmaceutical Assays, in the estimation of Impurities under Authorized Standards, and in Forensic Examinations for Poisons, with Directions for Elementary Organic Analysis. Third edition. 8vo, cloth $5.00

—— **Outlines of Proximate Organic Analysis, for the** Identification, Separation, and Quantitative Determination of the more commonly occurring Organic Compounds. 4th ed. 12mo, cl. $1.75

—— **First Book in Qualitative Chemistry. Eighth** edition. 12mo, cloth...$1.50

—— **and OTIS COE JOHNSON. Qualitative Chemical** Analysis. A Guide in the Practical Study of Chemistry and in the work of Analysis. Fourth fully revised edition. With Descriptive Chemistry extended throughout......................... $3.50

PRITCHARD, O. G. The Manufacture of Electric Light Carbons. Illustrated. 8vo, paper.................... .60

PULSIFER, W. H. Notes for a History of Lead. 8vo, cloth, gilt tops..$4.00

PYNCHON, Prof. T. R. Introduction to Chemical Physics, designed for the use of Academies, Colleges, and High Schools. Illustrated with numerous engravings, and containing copious experiments with directions for preparing them. New edition, revised and enlarged, and illustrated by 269 illustrations on wood. Crown 8vo, cloth...$3.00

RAFTER, GEO. W. and M. N. BAKER. Sewage Disposal in the United States. Illustrations and folding plates. Second edition. 8vo., Cloth....... .... $6.00

RAM, GILBERT S. The Incandescent Lamp and its Manufacture. 8vo., cloth.................................$6.00

RANDALL, J. E. A Practical Treatise on the Incan- descent Lamp. Illustrated. 16mo, cloth.................... .50

- — **P. M. Quartz Operator's Hand-book. New edition,** revised and enlarged, fully illustrated. 12mo, cloth...........$2.00

RANKINE, W. J. MACQUORN, C.E., LL.D., F. R. S. Applied Mechanics. Comprising the Principles of Statics and Cinematics, and Theory of Structures, Mechanism, and Machines. With numerous diagrams. Fourteenth edition. Thoroughly revised by W. J. Millar. Crown 8vo, cloth............................ $5.00

—— **Civil Engineering. Comprising Engineering Sur-** veys, Earthwork, Foundations, Masonry, Carpentry, Metal-Work, Roads, Railways, Canals, Rivers, Water-Works, Harbors, etc. With numerous tables and illustrations. Eighteenth edition. Thoroughly revised by W. J. Millar. Crown, 8vo, cloth.................... $6.50

RANKINE, W. J. MACQUORN, C.E., LL.D., F. R. S.
Machinery and Millwork. Comprising the Geometry, Motions, Work, Strength, Construction, and Objects of Machines, etc. Illustrated with nearly 300 woodcuts. Sixth edition. Thoroughly revised by W. J. Millar. Crown 8vo, cloth........ ....$5.00

—— **The Steam-Engine and Other Prime Movers. With** diagram of the Mechanical Properties of Steam, folding plates, numerous tables and illustrations. Thirteenth edition. Thoroughly revised by W. J. Millar. Crown, 8vo, cloth...................$5.00

—— **Useful Rules and Tables for Engineers and Others.** With appendix, tables, tests, and formulæ for the use of Electrical Engineers. Comprising Submarine Electrical Engineering, Electric Lighting, and Transmission of Power. By Andrew Jamieson, C.E., F.R.S.E. Seventh edition. Thoroughly revised by W. J. Millar. Crown 8vo, cloth...$4.00

—— **A Mechanical Text Book.** By Prof. Macquorn Rankine and E. F.B amber, C. E. With numerous illustrations. Fourth edition. Crown, 8vo, cloth...................................$3.50

RECKENZAUN, A. Electric Traction on Railways and Tramways. 213 Illustrations, 12mo, cloth..............$4.00

REED'S ENGINEERS' HAND-BOOK, to the Local Marine Board Examinations for Certificates of Competency as First and Second Class Engineers. By W. H. Thorn. With the answers to the Elementary Questions. Illustrated by 297 diagrams and 36 large plates. 15th edition, revised and enlarged. 8vo, cloth.........$4.50

—— **Key to the Fifteenth Edition of Reed's Engineers'** Hand-book to the Board of Trade Examinations for First and Second Class Engineers and containing the working of all the questions given in the examination papers. By W. H. Thorn. 8vo., cloth...... 3.00

RICE, Prof. J. M., and JOHNSON, Prof. W. W. On a New Method of obtaining the Differential of Functions, with especial reference to the Newtonian Conception of Rates or Velocities. 12mo, paper.................... .... .50

RIPPER, WILLIAM. A Course of Instruction in Machine Drawing and Design for Technical Schools and Engineer Students. With 52 plates and numerous explanatory engravings. Folio, cloth...$7.50

ROEBLING, J. A. Long and Short Span Railway Bridges. Illustrated with large copperplate engravings of plans and views. Imperial folio, cloth.............................$25.00

ROGERS, Prof. H. D. The Geology of Pennsylvania. A Government Survey, with a General View of the Geology of the United States, essays on the Coal Formation and its Fossils, and a description of the Coal Fields of North America and Great Britain. Illustrated with plates and engravings in the text. 3 vols, 4to, cloth, with portfolio of maps...................................$15.00

ROSE, JOSHUA, M.E. **The Pattern-Makers' Assistant.** Embracing Lathe Work, Branch Work, Core Work, Sweep Work, and Practical Gear Constructions, the Preparation and Use of Tools, together with a large collection of useful and valuable Tables. Seventh edition. Illustrated with 250 engravings. 8vo, cloth.....$2.50

—— **Key to Engines and Engine-Running.** A Practical Treatise upon the Management of Steam Engines and Boilers for the Use of Those who Desire to Pass an Examination to Take Charge of an Engine or Boiler. With numerous illustrations, and Instructions Upon Engineers' Calculations, Indicators, Diagrams, Engine Adjustments, and other Valuable Information necessary for Engineers and Firemen. 12mo, cloth...............................$3.00

SABINE, ROBERT. **History and Progress of the** Electric Telegraph. With descriptions of some of the apparatus. Second edition, with additions. 12mo, cloth..................$1.25

SAELTZER, ALEX. **Treatise on Acoustics in connec-** tion with Ventilation. 12mo, cloth.................... ......$1.00

SALOMONS, Sir DAVID, M.A. **Electric-Light Instal-** lations. A Practical Handbook. Seventh edition, revised and enlarged with numerous illustrations. Vol. I., The management of Accumulators. 12mo, cloth....................................$1.50
Vol. II., Apparatus, 296 illustrations. 12mo., cloth.........$2.25
Vol. III., Applications, 12mo., cloth.......................$1.50

SAUNNIER, CLAUDIUS. **Watchmaker's Handbook.** A Workshop Companion for those engaged in Watchmaking and allied Mechanical Arts. Translated by J. Tripplin and E. Rigg. Second edition, revised and appendix. 12mo, cloth............$3.50

SCHELLEN, Dr. H. **Magneto-Electric and Dynamo-** Electric Machines : their Construction and Practical Application to Electric Lighting, and the Transmission of Power. Translated from the third German edition by N. S. Keith and Percy Neymann, Ph.D. With very large additions and notes relating to American Machines, by N. S. Keith. Vol. 1., with 353 illustrations. Second edition, $5.00

SCHUMANN, F. A Manual of Heating and Ventilation in its Practical Application, for the use of Engineers and Architects. Embracing a series of Tables and Formulæ for dimensions of heating, flow and return pipes for steam and hot-water boilers, flues, etc. 12mo, illustrated, full roan.......................................$1.50

—— **Formulas and Tables for Architects and Engineers** in calculating the strains and capacity of structures in Iron and Wood. 12mo, morocco, tucks.......................................$1.50

SCIENCE SERIES, The Van Nostrand. [See List, p. 27]

SCRIBNER, J. M. Engineers' and Mechanics' Com-
panion. Comprising United States Weights and Measures. Mensuration of Superfices and Solids, Tables of Squares and Cubes, Square and Cube Roots, Circumference and Areas of Circles, the Mechanical Powers, Centres of Gravity, Gravitation of Bodies, Pendulums, Specific Gravity of Bodies, Strength, Weight, and Crush of Materials, Water-Wheels, Hydrostatics, Hydraulics, Statics, Centres of Percussion and Gyration, Friction Heat, Tables of the Weight of Metals, Scantling, etc., Steam and the Steam-Engine. Twentieth edition, revised. 16mo, full morocco..............................$1.50

SEATON, A. E. A Manual of Marine Engineering.
Comprising the Designing, Construction and Working of Marine Machinery. With numerous tables and illustrations reduced from Working Drawings. Twelfth edition. Revised throughout, with an additional chapter on Water Tube Boilers. 8vo, cloth.........$6.00

——and ROUNTHWAITE, H. M. A Pocketbook of Ma-
rine Engineering Rules and Tables. For the use of Marine Engineers and Naval Architects, Designers, Draughtsmen, Superintendents, and all engaged in the design and construction of Marine Machinery, Naval and Mercantile. Pocket size. Leather, with diagrams........$3.00

SHIELDS, J. E. Notes on Engineering Construction.
Embracing Discussions of the Principles involved, and Descriptions of the Material employed in Tunnelling, Bridging, Canal and Road Building, etc. 12mo, cloth...... ....$1.50

SHREVE, S. H. A Treatise on the Strength of Bridges
and Roofs. Comprising the determination of Algebraic formulas for strains in Horizontal, Inclined or Rafter, Triangular, Bowstring, Lenticular, and other Trusses, from fixed and moving loads, with practical applications, and examples, for the use of Students and Engineers. 87 woodcut illustrations. Fourth edition. 8vo, cloth,...$3.50

SHUNK, W. F. The Field Engineer. A Handy Book
of practice in the Survey, Location, and Truck-work of Railroads, containing a large collection of Rules and Tables, original and selected, applicable to both the Standard and Narrow Gauge, and prepared with special reference to the wants of the young Engineer. Tenth edition. Revised and Enlarged. 12mo, morocco, tucks.......$2.50

SIMMS, F. W. A Treatise on the Principles and Prac-
tice of Levelling. Showing its application to purposes of Railway Engineering, and the Construction of Roads, etc. Revised and corrected, with the addition of Mr. Laws' Practical Examples for setting out Railway Curves. Illustrated. 8vo, cloth,................$2.50

——— Practical Tunnelling. Explaining in detail Set-
ting-out of the Work, Shaft-sinking, Sub-excavating, Timbering, etc., with cost of work. 8vo, cloth......................... ....

SLATER, J. W. Sewage Treatment, Purification, and Utilization. A Practical Manual for the Use of Corporations, Local Boards, Medical Officers of Health, Inspectors of Nuisances, Chemists, Manufacturers, Riparian Owners, Engineers, and Rate-payers. 12mo, cloth...$2.25

SMITH, ISAAC W., C.E. The Theory of Deflections and of Latitudes and Departures. With special applications to Curvilinear Surveys, for Alignments of Railway Tracks. Illustrated. 16mo, morocco, tucks................... ....$3.00

——GUSTAVUS W. Notes on Life Insurance. Theoretical and Practical. Third edition. Revised and Enlarged. 8vo, cloth...

SNELL, ALBION T. Electric Motive Power: The Transmission and Distribution of Electric Power by Continuous and Alternate Currents. With a Section on the Applications of Electricity to Mining Work. 8vo., cloth, illustrated...................$4.00

STAHL, A. W., and WOODS, A. T. Elementary Mechanism. A Text-Book for Students of Mechanical Engineering. Fourth edition. Enlarged. 12mo, cloth.......................$2.00

STALEY, CADY, and PIERSON, GEO. S. The Separate System of Sewerage: its Theory and Construction. 8vo, cloth. With maps, plates and illustrations. Second edition....$3.00

STEVENSON, DAVID, F.R.S.N. The Principles and Practice of Canal and River Engineering. Revised by his sons David Alan Stevenson, B. Sc., F.R.S.E., and Charles Alexander Stevenson, B. Sc., F.R.S.E., Civil Engineer. Third edition, with 17 plates. 8vo, cloth...$10.00

—— The Design and Construction of Harbors. A Treatise on Maritime Engineering. Third edition, with 24 plates. 8vo, cloth......... ....$9.00

STEWART, R. W. A Text Book of Light. Adapted to the Requirements of the Intermediate Science and Preliminary Scientific Examinations of the University of London, and also for General Use. Numerous Diagrams and Examples. 12mo, cloth, $1.00

—— A Text Book of Heat. Illustrated. 8vo, cloth....$1.00

—— A Text-Book of Magnetism and Electricity. 160 Illustrations and Numerous Examples. 12mo, cloth..........$1.00

STILES, AMOS. Tables for Field Engineers. Designed for use in the field. Tables containing all the functions of a one degree curve, from which a corresponding one can be found for any required degree. Also, Tables of Natural Sines and Tangents. 12mo, morocco, tucks..$2.00

STILLMAN, PAUL. **Steam-Engine Indicator and the** Improved Manometer Steam and Vacuum Gauges; their Utility and Application. New edition. 12mo, flexible cloth..............$1.00

STONE, General ROY. **New Roads and Road Laws in** the United States. 200 pages, with numerous illustrations. 12mo, cloth...$1.00

STONEY, B. D. **The Theory of Stresses in Girders** and Similar Structures. With observations on the application of Theory to Practice, and Tables of Strength, and other properties of Materials. New revised edition, with numerous additions on Graphic Statics, Pillars, Steel, Wind Pressure, Oscillating Stresses, Working Loads, Riveting, Strength and Tests of Materials. 8vo, 777 pages, 143 illustrations, and 5 folding plates.......... ....$12.50

STUART, B **How to become a Successful Engineer.** Being Hints to Youths intending to adopt the Profession. Sixth edition. 12mo, boards.................................. .... .50

—— **C. B., C. E.** **Lives and Works of Civil and Mil-** itary Engineers of America. With 10 steel-plate engravings. 8vo, cloth..$5.00

SWEET, S. H. **Special Report on Coal, showing its** Distribution, Classification, and Costs delivered over different routes to various points in the State of New York and the principal cities on the Atlantic Coast. With maps. 8vo, cloth..................$3.00

SWINTON, ALAN A. CAMPBELL. **The Elementary** Principle of Electric Lighting. Illustrated. 12mo, cloth...... .60

TEMPLETON, WM. **The Practical Mechanic's Work-** shop Companion. Comprising a great variety of the most usefu rules and formulæ in Mechanical Science, with numerous tables of practical data and calculated results facilitating mechanical operations. Revised and enlarged by W. S. Hutton. 12mo, morocco.......$2.00

THOM, CHAS., and WILLIS H. JONES. **Telegraphic** Connections: embracing Recent Methods in Quadruplex Telegraphy. Oblong, 8vo, cloth. 20 full page plates, some colored.......$1.50

THOMPSON, EDWARD P. **How to Make Inven-** tions; or, Inventing as a Science and an Art. A Practical Guide for Inventors. Second edition. 8vo, boards....................$1.00

Toothed Gearing. **A Practical Hand-Book for Offices** and Workshops. By a Foreman Patternmaker. 184 Illustrations. 12mo, cloth...$2.25

TREVERT, EDWARD. **How to build Dynamo-Electric** Machinery, embracing Theory Designing and the construction of Dynamos and Motors. With appendices on Field Magnet and Armature Winding, Management of Dynamos and Motors, and Useful Tables of Wire Gauges. Illustrated. 8vo, cloth $2.50

TREVERT, EDWARD. Electricity and its Recent
Applications. A practical Treatise for Students and Amateurs, with
an Illustrated Dictionary of Electrical Terms and Phrases. Illustrated. 12mo, cloth..... ....$2.00

TUCKER, Dr. J. H. A Manual of Sugar Analysis, in-
cluding the Applications in General of Analytical Methods to the
Sugar Industry. With an Introduction on the Chemistry of Cane
Sugar, Dextrose, Levulose, and Milk Sugar. 8vo, cloth, illustrated...$3.50

TUMLIRZ, DR. O. Potential and its Application to
the Explanation of Electric Phenomena, Popularly Treated. Translated from the German by D. Robertson. Ill. 12mo, cloth....$1.25

TUNNER, P. A. Treatise on Roll-Turning for the
Manufacture of Iron. Translated and adapted by John B. Pearse.
of the Pennsylvania Steel Works, with numerous engravings, woodcuts.
8vo, cloth, with folio atlas of plates........................$10.00

URQUHART, J. W. Electric Light Fitting. Embody-
ing Practical Notes on Installation Management. A Hand-book for
Working Electrical Engineers—with numerous illustrations. 12mo,
cloth...$2.00

—— **Electro-Plating. A Practical Hand Book on the**
Deposition of Copper, Silver, Nickel, Gold, Brass, Aluminium, Plat-
ininum, etc. Second edition. 12mo........................$2.00

—— **Electro-Typing. A Practical Manual, forming a**
New and Systematic Guide to the Reproduction and Multiplication
of Printing Surfaces, etc. 12mo..........................$2.00

—— **Dynamo Construction: a Practical Hand-Book for**
the Use of Engineer Constructors and Electricians in Charge, embracing Frame Work Building, Field Magnet and Armature Winding and
Grouping, Compounding, etc., with Examples of Leading English,
American and Continental Dynamos and Motors, with numerous illustrations. 12mo, cloth.......................... ....$3.00

—— **Electric Ship Lighting. A Hand-Book on the**
Practical Fitting and Running of Ship's Electrical Plant. For the Use
of Ship Owners and Builders, Marine Electricians and Sea Going
Engineers in Charge. Numerous illustrations. 12mo, cloth......$3.00

UNIVERSAL (The) TELEGRAPH CIPHER CODE.
Arranged for General Correspondence. 12mo, cloth..........$1.00

VAN HEURCK, Dr. HENRI. The Microscope, Its
Construction and Management, including Technique, Photo-Micro-
graphy and the Past and Future of the Microscope. English edition
re-edited and augmented by the Author from the fourth French
edition, and translated by Wynne E. Baxter, F.R.M.S. 3 Plates and
upwards of 250 Illustrations. Imperial 8vo, cloth............$7.00

VAN NOSTRAND'S Engineering Magazine. Complete sets, 1869 to 1886 inclusive.
Complete sets, in 35 vols., in cloth...........................$60.00
Complete sets, in 35 vols., in half morocco..................$100.00

VAN WAGENEN, T. F. Manual of Hydraulic Mining. For the Use of the Practical Miner. 18mo, cloth..............$1.00

WALKER, W. H. Screw Propulsion. Notes on Screw Propulsion, its Rise and History. 8vo, cloth.................. .75

—— **SYDNEY F.** Electrical Engineering in Our Homes and Workshops. A Practical Treatise on Auxiliary Electrical Apparatus. Third edition revised, with numerous illustrations..$2.00

WANKLYN, J. A. A Practical Treatise on the Examination of Milk and its Derivatives, Cream, Butter, and Cheese. 12mo, cloth..$1.00

—— Water Analysis. A Practical Treatise on the Examination of Potable Water. Ninth edition. 12mo, cloth.....$2.00

WARD, J. H. Steam for the Million. A Popular Treatise on Steam, and its application to the Useful Arts, especially to Navigation. 8vo, cloth.....................................$1.00

WARING, GEO. E., Jr. Sewerage and Land Drainage. Large Quarto Volume. Illustrated with wood-cuts in the text, and full-page and folding plates. Cloth. Third edition.........$6.00

—— Modern Methods of Sewage Disposal for Towns, Public Institutions and Isolated Houses. 260 pages. Illustrated cloth $2.00

—— How to Drain a House. Practical Information for Householders. New and enlarged edition. 12mo, cloth........$1.25

WATT, ALEXANDER. Electro-Deposition. A Practical Treatise on the Electrolysis of Gold, Silver, Copper, Nickel, and other Metals, with Descriptions of Voltaic Batteries, Magneto and Dynamo-Electric Machines, Thermopiles, and of the Materials and Processes used in every Department of the Art, and several chapters on Electro-Metallurgy. With numerous illustrations. Third edition, revised and corrected. Crown 8vo, 568 pages.................$3.50

—— Electro-Metallurgy Practically Treated. Tenth edition, considerably enlarged. 12mo, cloth.................$1.00

WEALE, JOHN. A Dictionary of Terms Used in Architecture, Building, Engineering, Mining, Metallurgy, Archaelogy, the Fine Arts, etc., with explanatory observations connected with applied Science and Art. Fifth edition, revised and corrected. 12mo, cloth... :........$2.50

Weale's Rudimentary Scientific Series, [see list, p. 30.]

WEBB, HERBERT LAWS. A Practical Guide to the Testing of Insulated Wires and Cables. Illustrated. 12mo, cloth...$1.00

—— **The Telephone Hand Book.** 128 illustrations. 146 pages. 16mo., cloth..................... ....$1.00

WEEKES, R. W. The Design of Alternate Current Transformers. Illustrated. 12mo, cloth $1.00

WEISBACH, JULIUS. A Manual of Theoretical Mechanics. Eighth American edition. Translated from the fourth augmented and improved German edition, with an Introduction to the Calculus by Eckley B. Coxe, A.M., Mining Engineer. 1,100 pages, and 902 woodcut illustrations. 8vo, cloth...........$10.00 Sheep.. $11.00

WEYMOUTH, F. MARTEN.   Drum Armatures and Commutators. (Theory and Practice.) A complete Treatise on the Theory and Construction of Drum Winding, and of commutators for closed-coil armatures, together with a full resume of some of the principal points involved in their design, and an exposition of armature re-actions and sparking. 8vo, cloth $3.00

WEYRAUCH, J. J.   Strength and Calculations of Dimensions of Iron and Steel Construction, with reference to the Latest Experiments. 12mo, cloth, plates.....................$1.00

WHIPPLE, S., C.E.   An Elementary and Practical Treatise on Bridge Building. 8vo, cloth......................$4.00

WILKINSON, H. D. Submarine Cable-Laying, Re- pairing and Testing.....................................(In press)

WILLIAMSON, R. S. On the Use of the Barometer on Surveys and Reconnoissances. Part I. Meteorology in its Connection with Hypsometry. Part II. Barometric Hypsometry. With Illustrative tables and engravings. 4to, cloth..................$15.00

WILLIAMSON, R. S.   Practical Tables in Meteor- ology and Hypsometry, in connection with the use of the Barometer. 4to, cloth..$2.50

WILSON, GEO. Inorganic Chemistry, with New No- tation. Revised and enlarged by H. G. Madan. New edition. 12mo, cloth..$2.00

WOODBURY, D. V. Treatise on the Various Elements of Stability in the Well-Proportioned Arch. 8vo, half morocco..$4.00

WRIGHT, T.W. Prof.   A Treatise on the Adjustment of Observations. With applications to Geodetic Work, and other Measures of Precision. 8vo, cloth.............................$4.00

—— **A Text Book of Mechanics for Colleges and Techni-** cal Schools. Second edition. 12mo, cloth..................$2.50

WYATT, JAMES.   The   Phosphates   of   America.
Where and How They Occur ; How They are Mined ; and What They
Cost.   With Practical Treatises on the Manufacture of Sulphuric
Acid, Acid Phosphates, Phosphoric Acid and Concentrated Super-
phosphates and Selected Methods of Chemical Analysis.   20 Full
Page Plates and numerous Illustrations.   8vo, cloth...........$4.00

WYNKOOP,   RICHARD.   Vessels   and   Voyages, as
Regulated by Federal Statutes and Treasury Instructions and Decis-
ions.   8vo, cloth $2.00

Catalogue *of the* Van Nostrand Science Series.

THEY are put up in a uniform, neat, and attractive form. 18mo, boards. Price 50 cents per volume. The subjects are of an eminently scientific character, and embrace a wide range of topics, and are amply illustrated when the subject demands.

No. 1. **CHIMNEYS FOR FURNACES AND STEAM-BOILERS.** By R. Armstrong, C.E. Third American edition, revised and partly rewritten, with an appendix on Theory of Chimney Draught, by F. E. Idell, M.E.

No. 2. **STEAM-BOILER EXPLOSIONS.** By Zerah Colburn. New edition, revised by Prof. R. H. Thurston.

No. 3. **PRACTICAL DESIGNING OF RETAINING-WALLS.** By Arthur Jacob, A.B. Second edition, revised, with additions by Prof. W. Cain.

No. 4. **PROPORTIONS OF PINS USED IN BRIDGES.** Second edition, with appendix. By Charles E. Bender, C.E.

No. 5. **VENTILATION OF BUILDINGS.** By W. F. Butler. Second edition, re-edited and enlarged by James L. Greenleaf, C.E.

No. 6. **ON THE DESIGNING AND CONSTRUCTION OF STORAGE RESERVOIRS.** By Arthur Jacob, A.B. Second edition, revised, with additions by E. Sherman Gould.

No. 7. **SURCHARGED AND DIFFERENT FORMS OF RE-TAINING-WALLS.** By James S. Tate, C.E.

No. 8. **A TREATISE ON THE COMPOUND ENGINE.** By John Turnbull, jun. Second edition, revised by Prof. S. W. Robinson.

No. 9. **A TREATISE ON FUEL.** By Arthur V. Abbott, C. E. Founded on the original treatise of C. William Siemens, D.C.L.

No. 10. **COMPOUND ENGINES.** Translated from the French of A. Mallet. Second edition, revised, with Results of American Practice, by Richard H. Buel, C.E.

No. 11. **THEORY OF ARCHES.** By Prof. W. Allan.

No. 12. **A THEORY OF VOUSSOIR ARCHES.** By Prof. W. E. Cain. Second edition, revised and enlarged. Illustrated.

No. 13. **GASES MET WITH IN COAL-MINES.** By J. J. Atkinson. Third edition, revised and enlarged by Edward H Williams, jun.

D. VAN NOSTRAND COMPANY'S

No. 59. RAILROAD ECONOMICS; OR, NOTES, WITH COM-MENTS. By S. W. Robinson, C.E.

No. 60. STRENGTH OF WROUGHT-IRON BRIDGE MEM-BERS. By S. W. Robinson, C.E.

No. 61. POTABLE WATER AND THE DIFFERENT METH-ODS OF DETECTING IMPURITIES. By Charles W. Folkard.

No. 62. THE THEORY OF THE GAS-ENGiNE. By Dugald Clerk. Second edition. With additional matter. Edited by F. E. Idell, M.E.

No. 63. HOUSE DRAINAGE AND SANITARY PLUMBING. By W. P. Gerhard. Sixth edition, revised.

No. 64. ELECTRO-MAGNETS. By Th.du Moncel. 2d revised edition.

No. 65. POCKET LOGARiTHMS TO FOUR PLACES OF DECI-MALS.

No. 66. DYNAMO-ELECTRIC MACHINERY. By S. P. Thompson. With notes by F. L. Pope. Third edition.

No. 67. HYDRAULIC TABLES BASED ON "KUTTER'S FORMULA." By P. J. Flynn.

No. 68. STEAM-HEATING. By Robert Briggs. Second edition, revised, with additions by A. R. Wolff.

No. 69. CHEMICAL PROBLEMS. By Prof. J. C. Foye. Second edition, revised and enlarged.

No. 70. EXPLOSIVES AND EXPLOSIVE COMPOUNDS. By M. Bertholet.

No. 71. DYNAMIC ELECTRICITY. By John Hopkinson, J. A. Schoolbred, and R. E. Day.

No. 72. TOPOGRAPHICAL SURVEYING. By George J. Specht, Prof. A. S. Hardy, John B. McMaster, and H. F. Walling.

No. 73. SYMBOLIC ALGEBRA; OR, THE ALGEBRA OF ALGEBRAIC NUMBERS. By Prof. W. Cain.

No. 74. TESTING MACHINES: THEIR HISTORY, CON-STRUCTION, AND USE. By Arthur V. Abbott.

No. 75. RECENT PROGRESS IN DYNAMO-ELECTRIC MA-CHINES. Being a Supplement to Dynamo-Electric Machinery. By Prof. Sylvanus P. Thompson.

No. 76. MODERN REPRODUCTIVE GRAPHIC PROCESSES. By Lieut. James S. Pettit, U.S.A.

No. 77. STADIA SURVEYING. The Theory of Stadia Measurements. By Arthur Winslow.

No. 78. THE STEAM-ENGINE INDICATOR, AND ITS USE. By W. B. Le Van.

No. 79. THE FIGURE OF THE EARTH. By Frank C. Roberts, C.E.

No. 80. HEALTHY FOUNDATIONS FOR HOUSES. By Glenn Brown.

www.ingramcontent.com/pod-product-compliance
Lightning Source LLC
Chambersburg PA
CBHW021709210326
41599CB00013B/1589